I0475337

EasyTerms™
Terminology Guidebook
for Botany

Copyright 2009, Ed Creager

This edition of EasyTerms is one in a series of simple-to-use, college-level terminology guidebooks.

Although these guidebooks were originally intended for college students, many High School students will also find them helpful as they prepare for college.

Other topics covered in existing or forthcoming editions:

- Anatomy & Physiology (Human)
- Biochemistry
- Biology
- Business Management
- Cell Biology

- Ecology
- Genetics
- Microbiology
- Nursing
- Psychology
- Zoology

EasyTerms can help support your educational advancement and can boost the vocabulary of almost anyone who reads it.

For more information on these and other publications, please visit the author's site:

www.ApplecreekBooks.weebly.com

and please note the author's "signature book" entitled,

"The Money-Saving Idea Book: Inside Tips for Starving Students, Frugal Seniors and Every Financial Suvivor."

Foreword

This Botany edition is a simple-to-use, college-level* terminology guidebook and is part of the EasyTerms reference series. In the book, terms are arranged alphabetically within appropriate topic areas. The complete index makes it easy to find any term and its definition.

* These books can also help High School students prepare so that, before they attend college, they'll already know a considerable amount of the terminology they'll need.

A substantial number of the terms defined here have additional definitions outside the scope of the subject being covered. More general definitions and additional meanings, if sought, are to be found in less specialized publications such as dictionaries and encyclopedias.

Please check the website of the author...

AppecreekBooks.weebly.com

for more information on other available books.

To buy the Biology edition of EasyTerms and receive a Preferred Customer discount of 25%, please go to www.tinyurl.com/bookonbiology, click on "Add to Cart," and enter the **discount code "5B7CA4SP"** during check-out.

To save 25% on your copy of "The Money Saving Idea Book," go to www.tinyurl.com/tmsib, click on "Add to Cart," and enter the **discount code "DVRQQK9E"** during check-out.

Important Notice:

EasyTerms™
Terminology Guidebook

Table of Contents

The terms that follow are divided into the topics shown below. The page number on which the topic begins is given. Within each topic, the terms are arranged alphabetically.

Introduction

1.   anatomy

The study of structure.

2.   autecology

Study of interaction of an individual or species with its environment.

3.   biology

The study of life.

4.   biosphere

An interconnected system over the earth's surface in which organisms exist.

5.   botany

The study of plants.

6.   cell

A basic functional unit of a living organism.

7.   cell theory

A theory stating that living things are composed of cells.

8.   colloid

Glue-like; a particle in a colloidal dispersion.

9.   colloidal dispersion

A state of matter with small particles suspended in a medium.

10.   community

A set of interacting organisms living in a given location.

11.   continuity of life

The transfer of life from parent to offspring.

12.   data

Observations from an experiment.

13.   deductive reasoning

Reasoning from a general statement to a specific case.

14.   dendrochronology

Tree ring study for historical and climatic information.

15.   development

Process of increasing in complexity.

16.   diversity

The quality of variation among a class, such as living organisms.

17.   ecology

The study of how living organisms relate to each other and to their environment.

18.   ecosystem

All the organisms in a natural setting and their physical environment.

19.   environment

Changeable surroundings around a living organism.

20.   ethnobotany

The study of plants as they relate to human culture.

21.   flora

Plants of a region, as contrasted with fauna (animals).

22. fungi

A group of organisms consisting mainly of filaments and unable to synthesize their own food.

23. growth

Ability to increase in size.

24. herbarium

Collection of preserved plants.

25. hydrolysis

The splitting of a molecule with the addition of water.

26. hypothesis

A possible answer to a question; a possible explanation for observations that can be used to predict future outcomes.

27. inductive reasoning

Development of a general statement from a collection of observations.

28. internal environment

The environment around cells, but within the body.

29. interstitial

Concerning spaces between cells.

30. ionic bond

A chemical bond with atoms held together by the attraction of unlike charges.

31. kingdom

A major taxonomic subdivision of living organisms.

32. liter

The basic metric unit of fluid volume; 1.06 quart.

33. mass

The amount of matter in an object.

34. micrometer

Length equal to 1 millionth of a meter; micron.

35. microorganism

Organism not visible without magnification.

36. microscope

An instrument for observing structures too small to see with the naked eye.

37. millimeter

Length equal to one-thousandth of a meter.

38. mixture

Two or more substances combined in any proportions and retaining their individual properties.

39. morphology

Study of shape and appearance of organisms.

40. mycology

Study of fungi.

41. nanometer

Length equal to one billionth of a meter.

42. natural selection

A process by which well adapted organisms survive and reproduce in larger numbers than less well adapted ones.

43. nucleus

Central part of an atom or a cell.

44.   **organ**

A structure made up of several tissues that carries out particular functions; component of a system.

45.   **organism**

A living thing.

46.   **organizational complexity**

A concept that concerns the structural levels of an organism.

47.   **paleobotany**

Study of fossil plants and their relationship to the environment when they lived.

48.   **palynology**

Study of living or fossil pollen and spores.

49.   **phycology**

Study of algae.

50.   **physiology**

The study of life functions.

51.   **phytogeography**

Study of distribution of plants.

52.   **plant**

Member of the plant kingdom.

53.   **plant pathology**

Study of diseases of plants and their causes and control.

54.   **plant physiology**

Study of plant function.

55.   pteridology

Study of ferns.

56.   radiation

Spreading from a center; giving off electromagnetic particles and waves.

57.   reduction

Gain of an electron or loss of oxygen in a chemical reaction.

58.   reproduction

Process by which offspring arise.

59.   responsiveness

Ability to react to a stimulus.

60.   science

Study of natural laws by experimentation and hypothesis testing.

61.   scientific method

A method of collecting and testing data pertaining to a scientific question.

62.   secretion

A cell product; the active transport of substances from the blood to the kidney filtrate.

63.   synecology

Study of form and location of communities in relation to the environment.

64.   theory

An explanation that accounts for many observations.

65.   tissue

A group of similar cells, including intercellular substances, that carry out a particular function.

66. **vegetation**

All the plant life in a region.

67. **weight**

The amount of gravitational force exerted on an object.

Basic Chemistry

68.   acid

An ionizing substance that donates hydrogen ions.

69.   alkaline

Basic, able to accept hydrogen ions.

70.   anion

A negatively charge ion.

71.   atom

Smallest particle that retains properties of an element.

72.   atomic number

The number of protons in the nucleus of an atom.

73.   atomic weight

The total number of protons and neutrons in an atom; the average number if there are isotopes of the element.

74.   base

An ionizing substance that accepts hydrogen ions or reacts with an acid to form a salt.

75.   buffer

A substance that resists pH change by holding or releasing hydrogen ions in a solution.

76.   catalyst

A substance that increases a chemical reaction rate.

77.   cation

A positively charged ion.

78. compound

A substance with two or more elements combined in definite proportion.

79. covalent bond

A chemical bond formed by shared electrons between two atoms.

80. dehydration

Removal of water.

81. denaturation

An alteration in the shape and properties of a protein molecule.

82. electron

A negatively charged particle that continually moves around the nucleus of an atom.

83. element

A fundamental unit of matter.

84. endergonic

Requiring energy, as in a chemical reaction.

85. entropy

Tendency toward chaos or disorder.

86. exergonic

Releasing energy, as in a chemical reaction.

87. gram molecular weight

The quantity of a substance (in grams) equal to its molecular weight.

88. hydrogen bond

Weak covalent bond between hydrogen and another element, such as oxygen or nitrogen.

89. hydrophilic

Attacted to water.

90. hydrophobic

Tending to avoid water.

91. ion

A charged atom or group of atoms.

92. isomer

A molecule having the same kinds and number of atoms as another molecule, but arranged differently.

93. isotope

An atom having a different number of neutrons than certain other atoms of the same element.

94. kinetic

Pertaining to the energy of motion.

95. macromolecule

A very large molecule such as a protein or nucleic acid.

96. mole

A gram molecular weight.

97. molecule

The smallest quantity of a substance that retains its chemical properties.

98. neutron

An uncharged particle in the nucleus of an atom.

99. nonpolar

Lacking charged regions.

100. organic

Containing carbon.

101. oxidation

Addition of oxygen or loss of electrons in a chemical reaction.

102. polar

Having a positive or negative charge.

103. polar compound

A molecule having a charged area or polarity.

104. potential energy

Energy due to position and capable of being released, as in a rock at the top of a hill.

105. proton

A positively charged particle in the nucleus of an atom.

106. reactant

A substance that enters into a chemical reaction.

107. specific heat

The amount of heat needed to increase the temperature of a specific volume of substance one degree Celsius.

108. valence

An ion's charge.

Chemistry of Cells

109. adenosine triphosphate

An important energy storage molecule.

110. algin

A substance found in the cell wall of kelp and other algae and used in certain foods.

111. alkaloid

One of many nitrogenous organic bases found in seed plants.

112. amino acid

A molecule having both acid and amino functional groups.

113. autodigestion

Enzymatic digestion of spent parts of reproductive structures.

114. carbohydrate

An organic compound having several alcohol groups and an aldehyde or ketone group.

115. carrageenan

Jelly-like substance used to stabilize dairy products and made from certain red algae.

116. cellulose

A polysaccharide found in the structure of many plants.

117. chitin

Polysaccharide of glucose and nitrogenous units in the cell wall.

118. chitosan

Modified chitin in the cell walls of many Mucorales.

119. chromatography

Technique for separating substances according to their solubilities and color-producing reactions.

120. cicutoxin

Toxin from roots of water hemlock.

121. cutin

Waxy substance in cuticle.

122. deoxyribonucleic acid (DNA)

A nucleic acid in chromosomes that directs protein synthesis and transmits genetic information to a new generation.

123. disaccharide

A molecule having two sugar (saccharide) units held together by a glycosidic bond.

124. DNA

Deoxyribonucleic acid, cell's genetic material.

125. fatty acid

A long hydrocarbon chain with a carboxyl group at one end.

126. functional group

A component of a molecule that participates in a chemical reaction.

127. galactan

Polysaccharide containing glalactose units.

128. glucan

Polysaccharide consisting of glucose units.

129. glycine

An amino acid with the simplest chemical structure.

130. glycolipid

A molecule that contains both carbohydrate and lipid components.

131. glycoprotein

A molecule that contains both carbohydrate and protein components.

132. glycoside

Chemical substance produced from reaction between a sugar and another substance such as an alcohol.

133. hemicellulose

A kind of polysaccharide found in plant cell walls.

134. hexose

Six-carbon sugar.

135. hydrocarbon

Molecule consisting of carbon and hydrogen.

136. inorganic molecule

Molecule that lacks carbon.

137. lecithin

A phospholipid characteristic of animal tissues.

138. leucosin

Carbohydrate stored by golden-brown algae.

139. lignin

Complex polymer that strengthens cell walls.

140. lipid

Fat or fatlike substance.

141. lipoprotein

A molecule made of lipid and protein.

142. mannan

Polysaccharide consisting of mannose units in cell walls of some algae.

143. mannitol

Molecule consisting of mannose and an alcohol.

144. metabolism

All chemical reactions in a living organism.

145. monosaccharide

A simple sugar.

146. mucilage

Sticky, gelatinous substance from brown algae and some cycads.

147. nucleic acid

A polymer of nucleotides; DNA or RNA.

148. nucleotide

A molecule having a nitrogenous base, a 5-carbon sugar, and one or more phosphates.

149. oligosaccharide

Short polymer of sugar molecules.

150. organic molecule

Molecule containing carbon.

151. pentose

Five-carbon sugar.

152. peptide bond

A chemical bond between the amino group of one amino acid and the carboxyl group of another.

153. pH

Scale used to measure acidity or alkalinity.

154. phospholipid

A lipid made of glycerol, fatty acids, and phosphoric acid.

155. pigment

Molecule that can absorb some wavelengths of light.

156. pitch

A dark resin.

157. polymer

A molecule consisting of repeating units.

158. polypeptide

A chain of amino acids held together by peptide bonds.

159. polysaccharide

A molecule consisting of many saccharide units connected by glycosidic bonds.

160. protein

A polymer of amino acids.

161. resin

Any of many substances in pine and some tropical trees used in products such as medicines, wine, and varnish.

162. ribonucleic acid (RNA)

A nucleic acid made from information in DNA that is involved in protein synthesis.

163. RNA

Ribonucleic acid, nucleotide synthesized from DNA template.

164. rosin

Amber or dark colored translucent substance derived from resin.

165. saturated fatty acid

Fatty acid in which carbons lack double bonds and have maximum number of hydrogen substituents.

166. saturation

Condition of having all chemical affinities satisfied.

167. specificity

The attribute of being specific.

168. starch

Glucose polymer stored as reserve food in plants and some algae and fungi.

169. stereoisomer

Compound having the same kind and number of atoms as another compound, but in a different spatial arrangement.

170. steroid

A lipid with a complex four-ring structure.

171. substrate

Molecule upon which an enzyme acts.

172. sugar

Simple carbohydrate molecule produced by photosynthesis.

173. synthesis

Making of a substance.

174. tannin

Substance from oak, hemlock, and other barks used in tanning leather and making medicines and antimicrobials.

175. triose

Three-carbon sugar.

176. unsaturated fatty acid

Fatty acid with pairs of hydrogen atoms replaced by double bonds in the carbon chain.

177. uridine triphosphate (UTP)

A high energy molecule.

178. wax

Solid fat that waterproofs outer surfaces of stems and leaves.

Cells and Organelles

179. amylopectin

Storage polysaccharide; branching factor of starch.

180. amyloplast

Plastid that stores starch.

181. anthocyanin

Water-soluble pigment of different colors in cell vacuole.

182. basal body

Cylindrical structure having nine units of microtubules at the base of a flagellum or cilium.

183. cell culture

Laboratory procedure for growing cells in medium of known composition.

184. centriole

One of a pair of intracellular bodies that participate in forming a mitotic spindle.

185. chloroplast

A membrane-bound organelle containing chlorophyll and enzymes needed for photosynthesis.

186. chromatin

Nuclear material that condenses into distinct chromosomes during cell division.

187. chromoplast

Plastid containing carotene and similar pigments.

188. chromosome

In a human cell, one of 46 nuclear structures made of DNA and protein.

189. cilium

One of many hair-like process with characteristic pattern of microtubules that helps certain cells move.

190. cisterna

Circular, flat, membrane-bound sac.

191. coenzyme

A substance that works with an enzyme in activating chemical reactions.

192. concentration gradient

Range of concentrations of a substance over a region or across a membrane.

193. crista

Infolds in the inner mitochondrial membrane that are the site of electron transport.

194. cytokinin

A plant hormone associated with cell division.

195. cytology

Study of structure and function of cells.

196. cytoplasm

Cell substance, excluding the nucleus.

197. cytoplasmic streaming

Movement of chloroplasts and other structures within a cell's cytoplasm.

198. cytoskeleton

The organelles forming a cell's internal framework.

199. cytosol

The fluid part of cytoplasm that suspends organelles.

200. differential centrifugation

Technique that separates cell components by size and density.

201. electron micrograph

Image made with an electron microscope.

202. electrophoresis

Technique that separates charged molecules by subjecting them to a charged field.

203. endoplasmic reticulum

A membranous vesicular network within a cell.

204. eukaryote

Organism with membrane-bound organelles and nucleus in its cells.

205. eukaryotic

Having a nucleus and membrane-bound organelles.

206. exoenzyme

Enzyme secreted by a cell and acting to digest nutrients for the cell.

207. extracellular

Outside a cell.

208. extranuclear

Outside the nuclear envelope.

209. fixing

Killing and preserving cells and cell structure.

210. flagellum

Long, whiplike external structure with internal paired bundles of filaments.

211. free water

Water molecules not associated with solute molecules.

212. gas vacuole

Array of vesicles in prokaryotic cells.

213. gel

A semi-solid state in a colloidal dispersion.

214. glyoxysome

A microbody in a cell that functions in lipid metabolism.

215. Golgi apparatus

Membranous vesicles clustered in cells that complete synthesis of secretions.

216. heterokont

Having flagella of unequal length.

217. intracellular

Within a cell.

218. kinetochore

Region at which chromatids of a chromosome are connected.

219. leukoplast

Colorless plastid, usually storing starch.

220. lysosome

Membrane-bound organelle that contains digestive enzymes.

221. matrix

Central fluid-filled region of mitochondrion where Krebs cycle reactions occur.

222. microbody

Small membrane-bound organelle such as a lysosome.

223. microfibril

Structure consisting of cellulose polymers.

224. microfilament

A small, hollow protein fiber in cytoplasm that aids in movement or forms part of a cytoskeleton.

225. microtubule

Tubular structure made of the protein tubulin found in spindle fibers, flagella, and cilia.

226. mitochondrion

An organelle that contains enzymes for oxidative and energy-capturing processes.

227. monokaryotic

Having a single nucleus per cell.

228. nuclear

Of the nucleus.

229. nuclear envelope

Double membrane derived from endoplasmic reticulum that surrounds a nucleus.

230. nucleolar organizing region

Portion of certain chromosomes devoted to formation of nucleoli.

231. nucleolus

A body containing RNA within a nucleus.

232. nucleoplasm

The substance of a nucleus.

233. organelle

Small structure inside a cell having a particular function.

234. paramylum

Carbohydrate stored as food by euglenoids.

235. pectin

Chemical substance deposited in middle lamella that holds adjacent cells together.

236. peroxisome

An organelle containing oxidative enzymes.

237. plastid

One of a group of membrane-bound organelles capable of making starch.

238. prokaryotic

Lacking a nucleus and membrane-bound organelles.

239. proplastid

Immature plastid precursor of all plastids.

240. protoplasm

Cell substance; literally, first formed.

241. pyrenoid

Structure in a chloroplast where starch is deposited in many algae.

242. ribosome

An organelle containing ribonucleic acid and protein where protein synthesis occurs.

243. smooth ER

Endoplasmic reticulum lacking ribosomes where lipids are synthesized and secretory substances packaged.

244. sol

A liquid state of a colloidal dispersion.

245. stroma

Nonmembranous part of chloroplast where dark reactions occur.

246. surface-to-volume ratio

The surface area of a structure divided by its volume.

247. tonoplast

Membrane surrounding a vacuole.

248. tubulin

A protein that forms intracellular microtubules.

249. ultrastructure

Structure visible only at magnifications possible with electron microscopes.

250. vacuole

Membrane-bound structure containing water and metabolic by-products.

251. xanthophyll

Yellow, carotenoid pigment.

Membranes, Walls, & Movements

252. active transport

Transport of a substance against a gradient using a carrier molecule, enzyme, and cellular energy.

253. adsorptive endocytosis

Entry of a substance into a cell after attaching to the cell membrane.

254. binding site

A site where a particular molecule binds to a membrane or other structure.

255. carrier molecule

Membrane-associated molecule that helps transport particular substances across a membrane.

256. cell membrane

Lipid and protein compounds that form the boundary of a cell.

257. cell wall

Rigid outer boundary of a cell composed of cellulose, hemi-cellulose, and pectin.

258. cytosis

Evagination or invagination of cell membranes.

259. differential permeability

Variations in permeability of membranes to different substances.

260. diffusion

Net movement of molecules from higher to lower concentrations.

261. facilitated diffusion

Diffusion down a gradient on a carrier molecule but not requiring cellular energy.

262. filtration

Passage of a fluid across a membrane by mechanical pressure.

263. flaccid

Wilted because of water loss from cells.

264. fluid-mosaic model

Model showing arrangement of phospholipid bilayer with proteins interspersed in and on a cell membrane.

265. free space

Intercellular spaces and cell walls of a tissue.

266. gradient

The rate of change in the magnitude of concentration, pressure, or other variable.

267. hyperosmotic

Having higher osmotic pressure than a reference solution.

268. hypertonic

Causing movement of water out of cells.

269. hyposmotic

Having lower osmotic pressure than a reference solution.

270. hypotonic

Causing movement of water into cells.

271. isosmotic

Having the same osmotic pressure as a reference solution.

272. isotonic

Causing no net water movement across a cell membrane.

273. ligand

That which binds to a receptor.

274. lipid bilayer

Arrangement of two layers of lipids with polar heads toward surfaces and nonpolar tails associated between layers.

275. membrane

Phospholipid bilayer with interspersed proteins that regulates substances into and out of cells or organelles.

276. middle lamella

Structure that holds primary walls of adjacent structures together.

277. osmolarity

A solution's osmotic concentration determined by the number of osmotically active particles it contains.

278. osmosis

Diffusion of water through a membrane from its own higher to a lower concentration.

279. osmotic equilibrium

State in which no net movement of water occurs across a membrane.

280. osmotic pressure

Pressure created by osmosis.

281. passive transport

A process that moves substances without energy expenditure by the organism.

282. permeability

Capacity to penetrate.

283. pit

Small opening in secondary wall of a plant cell.

284. pit canal

Opening between a cell lumen and the pit chamber.

285. pit cavity

Space from the lumen of the cell to the pit membrane.

286. pit chamber

Opening in a bordered pit between the pit membrane and the outer aperture.

287. pit membrane

Segment of primary wall lying between pairs of pits in adjacent cells.

288. plasma membrane

Membrane forming the boundary of a cell.

289. plasmalemma

Outer cytoplasm that functions as a differentially permeable membrane.

290. plasmodesma

Intercellular cytoplasmic bridge that passes through cell walls to connect two adjacent protoplasts.

291. plasmolysis

The process by which a plant cell protoplast shrinks from its cell wall due to water loss in a hypertonic solution.

292. receptor

A specific site with which a specific substance can bind; cell that responds to signals sensed from the environment.

293. selectively permeable

A membrane property that allows passage of some substances while preventing passage of others.

294. sheath

Structure forming external covering around a cell wall.

295. sodium-potassium pump

Mechanism that actively moves Na ions out of cells and K ions into them against gradients.

296. solute

A dissolved substance.

297. solution

A liquid containing dissolved substances.

298. solvent

A substance in which other substances can dissolve.

299. suberin

Wax made by cork and endodermal cells and deposited in cell walls.

300. suberized

Containing suberin.

301. surface tension

Resistance to rupture by the surface film of a liquid.

302. tonicity

The degree to which fluid can move into or out of cells.

303. turgor pressure

Pressure created in plant cells by uptake of water from a hypotonic solution.

304. unit membrane

Membrane consisting of lipid bilayer with associated proteins.

Plant Forms

305. acellular

Having a thallus in which nuclei and other organelles are not compartmentalized.

306. acidophile

Plant able to live and thrive in acidic conditions.

307. anastomosis

Interconnections, as in leaf veins.

308. annual

Plant that completes life cycle in one year.

309. aphyllous

Lacking leaves.

310. arborescent

Looking like a tree.

311. biennial

Plant that completes its life cycle in two growing seasons.

312. bole

Lower portion of the trunk of a tree.

313. dendroid

Tree-like.

314. emergent tree

A tree that stands taller than the others in a forest.

315. endophyte

Plant growing within another plant.

316. epiphyte

A plant that grows on another plant.

317. evergreen

Plant that does not loose all leaves at the same time.

318. halophyte

Plant that grows in soil impregnated with salt.

319. hydrophyte

Plant having winter buds under water and usually perennial and aquatic.

320. insectivore

A plant that increases its nitrogen intake by capturing and ingesting insects.

321. leaf gall

Tumor on a leaf induced by certain insect eggs deposited on leaf.

322. macrophyte

Plant in a body of water large enough to see without a microscope.

323. mesophyte

Plant that grows in a moderately moist environment.

324. multicellular

Having many cells.

325. perennial

Plant that lives for many years, with living parts replacing those that die off each year.

326. saprophyte

An organism that feeds by absorbing molecules from decaying organic matter.

327. seedling

Immature plant that developed from a seed.

328. shrub

Woody plant with two or more stems arising near ground.

329. sporeling

Immature plant produced from a spore.

330. succulent

Plant that stores water in stems or leaves.

331. thermophile

Organism that grows well at high temperature.

332. tree

Woody plant with trunk (main stem) that produces branches some distance above ground.

333. unicellular

Having a single cell.

334. xerophyte

Plant that can grow under dry conditions.

Plant Descriptors

335. abaxial

Away from the axis of a plant.

336. abiotic

In the absence of life.

337. adaxial

Toward the axis of a plant.

338. aerial

Above ground.

339. amorphous

Lacking any particular form.

340. arid

Extremely dry.

341. axial

Of or in the direction of the axis of a plant.

342. calcareous

Containing or consisting of calcium carbonate.

343. coralline

Describing certain lime-crusted red algae.

344. crenulate

Having a scalloped shape.

345. cross-section

Section through a structure perpendicular to a long axis.

346. cruciate

Having a cross shape.

347. crustose

Having a thin thallus closely attached to rock, soil, or bark.

348. deciduous

The quality of losing leaves in winter or drought.

349. decussate

Occuring alternately at right angles.

350. desiccation

Drying out.

351. dimorphic

Capable of assimilating nutrients by either hyphae or budding cells, as in some fungi.

352. distal

Most distant from the point of origin of a structure.

353. distromatic

Having a thallus with two layers of cells.

354. endogenous

On the inside.

355. ephemeral

Transient, temporary.

356. equitant

Overlapping property of leaves in plants such as irises.

357. exogenous

On the outside.

358. external

On the outside.

359. fascicular

Contained within a bundle, as vascular cambium in vascular bundle.

360. foliaceous

Leaflike.

361. foliose

Leafy, as in some lichens.

362. fructicose

Shrub-like, branched form, as in a lichen.

363. frugivorous

Concerning a fruit-eating organism.

364. fusiform

Spindle-shaped.

365. helical

Spiral-shaped.

366. intercalary

Interposed between two cells or tissues.

367. interfascicular

Located in space between two vascular bundles.

368. internal

Inside or within.

369. intrinsic

Entirely within.

370. laminate

Broadened and flattened.

371. lanceolate

Tapering at the ends.

372. lateral

On or toward the side.

373. lumen

Space within a dead cell.

374. monopodial

Having a single main axis of growth.

375. nonlignified

Lacking lignin.

376. ostiole

Pore or opening.

377. palmate

Hand-like.

378. peripheral

Outer or away from the center.

379. planated

Having dichotomous branches arranged in a single plane.

380. pleomorphism

Having more than one normal shape.

381. polarity

Having different form and function at opposite ends of a cell, tissue, or organ.

382. polystromatic

Having many cell layers.

383. proximal

Nearest the point of origin of a structure.

384. pyriform

Pear-shaped.

385. radial

Plane parallel to long axis and passing through center.

386. radial section

Lengthwise section passing through center of object.

387. radial symmetry

Divisible into equal halves by any line along a radius.

388. reticulate

Net-like.

389. reticulum

A mesh or network.

390. sclerified

Hardened because of lignin deposition in secondary wall.

391. septate

Partitioned.

392. septum

Crosswall that forms between cells during cell division.

393. serrate

Having toothed margins.

394. spatulate

Spoon-like.

395. squamulose

A foliose growth with loosely attached thallus lobes.

396. subtending

Extending beneath.

397. tangential section

Lengthwise section that does not pass through the center.

398. temperate

Having a moderate climate.

399. toothed

Having toothlike indentations along margin of a leaf blade.

400. transverse

Crosswise.

401. truncate

Cut off.

402. umbonate

Having a convex hump at the center.

403. woody

Having large amounts of secondary xylem.

404. xeric

Dry.

Genetics and Mutations

405. allele

One of several different genes for a trait that can occupy a particular site on a chromosome.

406. allopolyploidy

Multiple sets of chromosomes in a hybrid that had few if any homologous chromosomes.

407. allotetraploidy

Having four sets of chromosomes from replication of two different genomes from a cross.

408. aneuploidy

Having more or less than the normal diploid number of chromosomes.

409. autoploidy

Polyploidy due to replication of a single genome.

410. autosomal

Concerning paired (nonsex) chromosomes and the genetic information they carry.

411. autosome

One of a pair of nonsex chromosomes.

412. auxotrophic

A mutant organism that has greater nutrional requirements than its normal counterpart.

413. classical genetics

Principles of inheritance that apply to observable characteristics.

414. codominance

The simultaneous expression of two alleles in one individual without blending.

41

415. crossing-over

The exchange of corresponding segments of DNA during meiosis.

416. deletion

Loss of one or more bases from a DNA strand.

417. dihybrid

Involving two pairs of alleles.

418. DNA hybridization study

Analysis of extent to which different single strands of DNA can hybridize (join) to make a double strand.

419. dominant allele

The allele expressed when it and a recessive allele are present.

420. epistatic

One gene interfering with the effect of another gene.

421. frameshift mutation

A DNA sequence change caused by adding or deleting bases.

422. gene

Functional unit of heredity; a site on a chromosome that transmits a particular hereditary characteristic.

423. gene amplification

Selective replication of certain genes.

424. gene sequencing

Determination of the order of nucleotides in a gene.

425. gene therapy

Biotechnical applications designed to treat genetic disease.

426. **genetic engineering**

Use of human-designed procedures to alter genetic information.

427. **genetic information**

Information coded in DNA that determines protein structure.

428. **genetic load**

The number of genetic defects present in a population.

429. **genetic material**

DNA molecules that contain coded information.

430. **genetic screening**

Search for genetic defects in fetuses, newborns, and prospective parents.

431. **genetics**

The study of heredity.

432. **genome**

An organism's whole complement of DNA.

433. **genomic library**

A collection of genetically engineered viruses carrying all the genes of a species.

434. **genotype**

Alleles of a single gene or all the genes carried by a particular individual.

435. **hemizygous**

Having one of each kind of sex chromosome.

436. **heredity**

Transmission of characteristics from one generation to the next.

437. heterosis

Vigorous growth capacity, especially as seen in crossbred plants.

438. heterozygous

Having unlike alleles for a trait.

439. homologous

Having the same shape and structure; pairs of chromosomes having the same information.

440. homozygous

Having like alleles for a trait.

441. hybrid

An offspring produced by crossing populations differing in one or more traits.

442. hybrid vigor

The display of increased fitness resulting from crossing of populations.

443. hybridization

Mixing of genes from different parental strains during sexual reproduction.

444. incomplete dominance

The expression by blending of two alleles present at the same time in an organism.

445. inheritance

Transmission of characteristics from one generation to the next.

446. insertion mutation

A change in DNA involving the addition of nucleotides at some point along the strand.

447. inversion

A change in a chromosome resulting in a reordering of its genes.

448. karyotype

Arrangement of chromosomes from a cell in pairs and in a fixed order.

449. linkage

Degree to which genes are closely associated physically and thereby inherited together.

450. linkage group

A group of genes physically close and inherited together.

451. locus

A particular site on a chromosome were a gene is located.

452. maternal inheritance

Inheirtance of characteristics from maternal parent only by way of genes in plastids and mitochondria.

453. missense mutation

A point mutation that replaces one amino acid for another in a protein.

454. molecular genetics

Study of inheritance at the molecular level.

455. molecular hybridization

A procedure for determining similarity of nucleotide sequence between two nucleic acid molecules.

456. monohybrid

Pertaining to a genetic cross that differs in a single trait under study.

457. mosaic

Organism with some cells containing different genetic information than other cells.

458. multiple alleles

Three or more kinds of genetic information for a given trait.

459. mutagen

An agent that can alter DNA.

460. mutation

A change in genetic information.

461. nonsense mutation

A point mutation that produces a codon that stops protein synthesis, thereby creating a short nonfunctional peptide.

462. outcrossing

Mating between strains of the same breed.

463. pentaploid

Having five of each kind of chromosome; 5N.

464. phenotype

The appearance of an individual with respect to one or all inherited characteristics.

465. phenotypic plasticity

Expression of different phenotypes from one genotype under different environmental conditions.

466. pleiotropy

The influence of a single gene on more than one trait.

467. ploidy

Number of sets of chromosomes.

468. point mutation

A change in a single base in a DNA molecule.

469. polygenic inheritance

A situation in which two or more genes, each with alleles, jointly affect the expression of a trait.

470. polyploidy

Presence of more than two sets of chromosomes.

471. principle of dominance

A Mendelian principle that one factor for a trait can mask or overpower another factor for the same trait.

472. principle of unit characters

A Mendelian principle that individuals carry two factors for each trait.

473. Punnett square

A table to illustrate the distribution of alleles in the offspring of heterozygotes.

474. recessive

In genetics, a characteristic seen in the phenotype only when recessive allele is only one present in the genotype.

475. recombinant DNA

DNA segments combined from two different organisms.

476. recombination

Combining of alleles from parents as a result of meisosis and sexual reproduction.

477. segregation

Separation of homologous chromosomes during meiosis.

478. selection

Favoring of certain alleles in a gene pool.

479. sex chromosome

A chromosome associated with maleness or femaleness; X or Y chromosome in mammals.

480. structural gene

A gene that produces a specific product.

481. teratogen

An agent that causes defective embryonic development.

482. testcross

The mating of an organism with an unknown genotype to one with a homozygous recessive phenotype.

483. tetrad

Four copies of the same chromosome temporarily attached to each other.

484. tetraploidy

Having four sets of chromosomes.

485. transduction

Genetic exchange in which DNA is transferred by a virus from one bacterium to another.

486. transformation

Genetic exchange in which DNA is transferred from one cell to another.

487. triploidy

Having three sets of chromosomes.

488. trisomy

Condition of having three copies of a chromosome.

489. variation

Divergence among individuals in a species.

490. variety

Subgroup within a species, usually from controlled interbreeding.

Protein Synthesis & Cell Division

491.   anaphase

A mitotic stage during which chromosomes move apart.

492.   anticodon

Three-base sequence on tRNA that fits with codon of mRNA.

493.   aster

Short microtubules at the ends of a spindle in a dividing animal cell.

494.   biotechnology

The use of a natural biological system to make a product or achieve a particular end.

495.   cell cycle

A repetitive sequence of events involving DNA replication and cell division.

496.   cell division

Process by which a mother cell gives rise to daughter cells.

497.   centromere

Constriction in a chromosome where chromatids are joined and spindle fibers attached.

498.   chromatid

One of two lengthwise halves of a replicated chromosome.

499.   codon

A three-base sequence in messenger RNA derived from DNA and specifying amino acid placement in a protein.

500.   complementary base pairing

Bonding between certain bases in nucleic acid strands.

501. **cytokinesis**

Division of the cytoplasm that follows division of a nucleus.

502. **DNA ligase**

An enzyme that attaches cut ends of DNA molecules.

503. **DNA polymerase**

An enzyme that increases chain length in DNA synthesis.

504. **DNA replication**

Synthesis of new DNA according to information in an existing DNA template.

505. **equational division**

Second meiotic division in which chromosomes do not replicate.

506. **free-nuclear division**

Development by repeated division of primary nucleus.

507. **G-1 phase**

Stage of cell cycle after cell division in which RNA and proteins are synthesized and the cell grows.

508. **G-2 phase**

Stage of cell cycle in which spindle fiber proteins are synthesized before division begins.

509. **genetic code**

The three-base sequences in messenger RNA derived from a DNA template that determine amino acid order in proteins.

510. **haploid**

Having one of a pair of chromosomes.

511. **inducer**

A regulatory molecule that promotes expression of a gene.

512. inducible enzyme

An enzyme synthesized only in the presence of its substrate.

513. interphase

A cell cycle stage during which the cell is not dividing.

514. lagging strand

The DNA strand on which synthesis is discontinuous during replication.

515. leading strand

The DNA strand that is replicated continuously.

516. messenger RNA

A nucleic acid that carries information in the form of codons for the synthesis of a protein.

517. metaphase

A mitotic stage during which chromosomes align along the equator of a cell.

518. mitosis

Nuclear division that produces two identical nuclei.

519. mitotic toxin

Chemical substance that dirupts mitosis.

520. Okazaki fragments

Short segments of single-stranded DNA synthesized on a lagging strand.

521. operon

A group of bacterial genes that function together and are controlled by a regulator gene.

522. plasmid

Extrachromosomal DNA that replicates independently in a host cell.

523. prophase

The first mitotic stage during which the chromosomes become distinct.

524. purine

A nitrogenous base with two rings found in nucleic acids.

525. pyrimidine

A nitrogenous base with one ring found in nucleic acids.

526. regulator gene

A gene that directs the activity of a specific set of genes.

527. replicated

Duplicated or copied.

528. replication

Duplication.

529. repressor

A gene that prevents the action of a set of genes.

530. restriction enzyme

An endonuclease that cuts double stranded DNA at sites having specific nucleotide sequences.

531. restriction site

The site at which an endonuclease acts.

532. reverse transcriptase

An enzyme that makes DNA according to an RNA template.

533. ribosomal RNA (rRNA)

A nucleic acid that forms part of a ribosome.

534. S-phase

Synthetic phase of cell cycle when DNA is replicated.

535. semiconservative replication

The replication of DNA in which each molecule consists of one new and one old strand.

536. spindle fiber

Microtubules in eukaryotic cells involved in the movement of chromosomes during mitosis and meiosis.

537. telophase

The last mitotic stage during which nuclei reform.

538. template

Pattern.

539. transcription

The transfer of coded genetic information from DNA to mRNA.

540. transfer RNA

Ribonucleic acid that binds and carries amino acids to a specific site in a growing peptide chain.

541. translation

The process by which mRNA codons are used to determine the sequence of amino acids in a protein.

542. translocation

Transfer of part of a chromosome from its normal location to a location on another chromosome.

543. vector

A DNA carrier that can insert foreign genetic material into a host cell.

Tissues

544. adaxial meristem

Tissue with dividing cells on adaxial side of leaf.

545. annular tracheid

A xylem cell having separate rings of lignocellulose in its secondary wall.

546. apical meristem

Undifferentiated cells that continuously divide at the tip of a stem or root.

547. bark

Tissues outside vascular cambium.

548. bud primordium

Cluster of meristematic cells that produce a bud.

549. bundle cap

A layer of fiber cells that surround and protect the primary phloem.

550. bundle sheath

A layer of parenchymal cells that surround and protect vascular tissues in a leaf or monocotyledonous stem.

551. callus

Thickened region on the surface of a plant.

552. cambium

A sheet of unspecialized cells between xylem and phloem that can divide to produce either tissue.

553. casparian strip

Continuous suberin band in the wall of an endodermal cell.

554. collenchyma

Supporting cells with reinforced walls found in elongating stems and petioles of plants.

555. complex tissue

Tissue containing more than one cell type.

556. cork

Secondary tissue made of suberized cells found on outer layer of roots and woody stems.

557. cork cambium

Secondary meristem that produces cork and phelloderm.

558. cuticle

Waxy outer layer of epidermis.

559. derivative

Cell derived from a meristematic cell.

560. dermal tissue

Tissue located on the outside of an organ.

561. detached meristem

A tissue with dividing cells separated by some distance from apical meristem.

562. dilated ray

V-shaped region of parenchymal cells in secondary phloem of woody plants.

563. endocarp

Inner layer of pericarp in certain fruits such as oranges.

564. endodermis

Single layer of thick-walled and thin-walled cells around the vascular cylinder of a root.

565. epidermis

Single layer of tissue on the outside of roots and stems of nonwoody plants and leaves.

566. fiber

Vertically elongated, dead cell with thick secondary wall.

567. filament

Cells linked end-to-end forming a chain.

568. ground meristem

A primary meristem that gives rise to ground tissues such as collenchyma and parenchyma.

569. ground tissue

Primary tissue derived from ground meristem.

570. guard cell

Bean-shaped epidermal cell containing chloroplasts that regulates opening and closing of a stoma.

571. herbaceous plant

Plant composed mainly of primary tissue.

572. herbaceous stem

A soft green stem with little woody tissue.

573. hypodermis

Cell layer immediately internal to epidermis.

574. intercalary meristem

Region of dividing cells located some distance from the apex.

575. lacuna

Air space in a tissue.

576. leaf primordium

Meristematic cells that give rise to a leaf.

577. marginal meristem

Region of dividing cells along the margin of a developing leaf.

578. meristem

Plant tissue containing cells capable of dividing throughout the life of the plant.

579. mucilage duct

Duct containing mucilage, gum, or another plant product.

580. mucilage hair

Small mucilage-producing projections from growing points of bryophytes.

581. palisade cell

Part of a compact layer of cells in the mesophyll of fern and seed plant leaves.

582. palisade parenchyma

Cells arranged in columns under upper leaf epidermis that carry on photosynthesis.

583. parenchyma

Cells that form packing material inside leaves, stems, and roots.

584. passage cell

Endodermal cell that allows passage of water into root xylem.

585. pericycle

Single-layered tissue that produces branch roots.

586. periderm

Protective tissue that replaces the epidermis in widening stems; cork, cork cambium, and cork parenchyma.

587. phelloderm

Parenchyma produced by cork cambium.

588. phellogen

Meristematic cells that produce cork; cork cambiuim.

589. phellum

Cork cells that produce the outer layers of periderm.

590. pinnule

Smallest subdivision of a compound leaf or frond.

591. pith

Tissue at the center of the stem or root of a plant.

592. pith ray

Strip of parenchyma connecting pith and cortex and extending between vascular bundles in a herbaceous stem.

593. plasmodesmata

Extensions of cytoplasm between adjacent cells that maintain intercellular communication and continuity.

594. plate meristem

Dividing cells between midrib of a leaf and marginal meristem.

595. primary meristem

Meristem that produces primary tissues.

596. primary phloem

Phloem formed from procambium.

597. primary tissue

Tissue composed of cells produced by meristem at the tips of roots and stems.

598. **primary wall**

First layer of cell wall formed by a new cell.

599. **primary xylem**

Xylem formed from procambium.

600. **procambium**

Primary meristem that forms primary vascular tissues and pericycle.

601. **prothallial cell**

Sterile cells produced along with pollen grains in nonflowering plants.

602. **protoderm**

Primary meristem that gives rise to epidermis.

603. **protophloem**

First phloem to mature.

604. **protoxylem**

First formed xylem.

605. **quiescent center**

Region of slower cell division in the tip of root meristem.

606. **ray**

Radially arranged spokes of mainly parenchyma arising from cambium and extending into secondary vascular tissues.

607. **ray tracheid**

Thick lignified cells with bordered pits at ends of rays in some coniferous woods.

608. **rhytidome**

Outer layer of tissue in bark.

609. sapwood

Outer part of xylem in a root or woody stem that is light in color and can transport minerals and water.

610. scale

A plate-like outgrowth in a vascular plant; a cell cover in some algae.

611. sclereid

Kind of dead cell with thick wall and pit canals that makes up sclerenchyma tissue.

612. sclerenchyma

A supporting tissue in thick-walled plant cells.

613. secondary meristem

Meristem that gives rise to a secondary tissue.

614. secondary tissue

Tissue originating from the vascular cambium or cork cambium and that increases root or stem diameter.

615. secondary wall

Layer of cell wall deposited on primary layer that usually results in death of cell.

616. secondary xylem

Conducting and supporting tissue from vascular cambium; wood.

617. simple tissue

Tissue having a single cell type.

618. spiral tracheid

Tracheid with spiral-shaped secondary wall.

619. spongy mesophyll

Aggregation of lobed cells with large spaces between them on the underside of a leaf.

620. **spongy parenchyma**

Photosynthetic tissue in a leaf which has loosely arranged cells.

621. **syngenesis**

Fusion or coming together of objects.

622. **thallus**

Undifferentiated body of a plant.

623. **tissue culture**

Procedure for growing tissues in the laboratory.

624. **tracheary element**

Conducting element of xylem.

625. **tracheid**

Kind of water-containing xylem cell with narrow diameter and end walls.

626. **transition region**

Region in a hypocotyl where transition between root and stem vascular tissues occurs.

627. **vascular cambium**

Secondary meristem that produces secondary xylem and phloem.

628. **vascular ray**

Radial strip of cells in secondary vascular tissue.

629. **vascular system**

A plant's support and conduction system consisting of xylem and phloem.

630. **vascular tissue**

Tissue that conducts nutrients or water through plants.

631. xylem

Tissue specialized to conduct water and dissolved minerals.

632. xylem ray

Parenchymal cells in xylem.

Roots, Stems, Minerals, Transport

633. absorption

Movement of substances across a membrane.

634. absorptive

Concerning absorption.

635. adventitious bud

Bud from tissue other than apical meristem.

636. adventitious root

Root from tissue other than a root.

637. aerial root

Above ground root that obtains moisture from air.

638. apical bud

Growing tip of a stem.

639. apical dominance

A terminal bud's inhibition of lateral bud growth.

640. assimilation

Taking in and absorbing into the body.

641. axil

Angle between a leaf and its stem.

642. axillary bud

Bud located in an axil.

643. bordered pit

A pit having its secondary wall overarching the pit membrane.

644. bordered pitted tracheid

A tracheid having bordered pits, usually on its radial wall.

645. branch

Stem that arises from an axillary bud.

646. branch root

Root formed as a branch of an existing root.

647. breathing root

Root of swamp plant that is exposed to air and can exchange gases.

648. bud

A short, immature section of a plant stem.

649. bud scale

Modified leaf that surrounds and protects a bud.

650. buttress root

Adventitious root on a stem that helps to support a plant.

651. capillary action

The ability of a liquid to rise against gravity because of molecular cohesiveness and adherence.

652. cladode

Modified stem that acts as photosynthetic organ.

653. climbing root

Adventitious root derived from a stem that attaches a plant to a vertical surface.

654. companion cell

A specialized cell found adjacent to a sieve-tube cell.

655. contractile root

A root that shortens during development, changing the position of the shoot relative to the ground.

656. cortex

Root or stem region made mainly of parenchyma cells between vascular cylinder and epidermis.

657. dictyostele

Stele having separate bundles of xylem and phloem.

658. diffuse root system

Root system of monocotyledonous plants where primary root is replaced by many smaller roots; fibrous root system.

659. ectomycorrhiza

Symbiotically related fungi and tree roots in which fungi grow around tree roots.

660. endomycorrhiza

Symbiotic relationship between fungi and roots with fungal filaments inside root tissue.

661. evapotranspiration rate

Rate at which water vapor is lost through stomata.

662. gas exchange

The diffusion of gases across membranes as when oxygen enters and carbon dioxide leaves blood.

663. ingestion

Intake of food or fluid.

664. internode

The space between nodes on a plant stem.

665. knee

Modified root on certain plants that grow in swamps.

666. lateral bud

Bud at axil that gives rise to a branch.

667. lenticel

A group of loosely arranged cork cells that allow gas exchange.

668. macronutrient

Nutrient needed in relatively large amounts.

669. mass/pressure flow hypothesis

Idea that nutrient movement depends on pressure differences between food-making and food-storing regions of phloem.

670. micronutrient

A nutrient needed in relatively small quantities.

671. mineral

Inorganic substance.

672. node

A portion of a stem that produces leaves or branches in stems.

673. nodule

Small spherical body on the root of a legume containing nitrogen-fixing bacteria.

674. nonsuberized

Lacking the wax suberin.

675. nutrition

The act of providing substances needed for good health through food ingestion.

676. petiole

A structure that connects a leaf to a stem.

677. phloem

The carbohydrate transporting tissue of a plant.

678. prop root

Root modified as main support for plant.

679. protostele

Stele with a solid column of vascular tissue.

680. ratoon

A shoot from a perennial plant.

681. rhizome

A horizontal underground stem.

682. root

Main underground organ of most plants, derived from radicle and used to anchor plant and absorb minerals and water.

683. root cap

Cells that protect growing tip of a root.

684. root hair

Small tubular process of a root epidermal cell that increases its absorptive area.

685. root hair zone

Segment of a root where root hairs are present.

686. root nodule

Spherical growth on root containing nitrogen-fixing bacteria and root tissue.

687. root pressure

Force caused by differences in osmotic pressure between the cells of a root hair and cells in xylem or root pith.

688. runner

A long narrow stem growing horizontally along the surface of the ground; stolon.

689. scion

In grafting, the stem cutting that is inserted into the stock.

690. shoot

Part of plant above ground consisting of stems, leaves, buds, and reproductive organs.

691. shoot system

All a plant's shoots.

692. sieve area

Concentration of pores in a sieve tube or sieve cell.

693. sieve cell

Food conducting cell of phloem in a gymnosperm.

694. sieve plate

Perforated end of cell in sieve tube.

695. sieve tube

Several sieve cells arranged end-to-end.

696. sieve tube element

One cell of several that form a sieve tube.

697. silica

Silicon dioxide, component of sand.

698. siphonostele

Stele consisting of central pith surrounded by a cylinder of vascular tissue.

699. sleep movement

Reversible nastic movement regulated by turgor pressure, as in folding of leaves at night.

700. stele

Part of a root or stem that contains vascular tissues and is surrounded by cortex.

701. stipe

Stalk that has no vascular tissue.

702. stolon

A long narrow stem growing along the ground; runner.

703. sucker

Modified root that produces new plants.

704. sympodium

Stem consisting of superimposed branches that function as an axis.

705. tap root system

Root system with single primary root that produces multiple secondary roots found in dicotyledonous plants.

706. tendril

Coillike leaf or stem that attaches a weak-stemmed plant to a support.

707. terminal-bud-scale scar

Circular mark indicating position of previous terminal bud.

708. thorn

Modified stem with hard spiny structure.

709. **transpiration**

Water evaporation from stoma of leaves and stems.

710. **transpirational-pull cohesion hypothesis**

Idea that water moves in xylem by pull created by transpiration on cohesive column of water.

711. **trunk**

Main tree stem.

712. **tuber**

A bulky terminal part of an underground stem.

713. **turgor**

Firmness in cells due to water held in them by osmotic pressure.

714. **twiner**

Weak-stemmed plant that requires support for upright growth.

715. **tyloses**

Blockage of tracheid by growth of parenchymal cell through a pit.

716. **vascular bundle**

Aggregate of xylem and phloem.

717. **vascular bundle scar**

Mark at position of vascular bundle within a leaf scar on a stem.

718. **vesicle**

Sac-like structure.

719. **vessel**

Group of vessel elements arranged end-to-end to make a tube of xylem tissue.

720.   vessel element

A cells comprising a vessel in xylem.

Leaves and Photosynthesis

721. abscission layer

Layer of cells that grow as leaf or fruit separates from stem.

722. alternate leaf arrangement

Placement of one leaf per node on alternating sides of stem.

723. autotroph

An organism that makes organic compounds from inorganic substances in the environment.

724. bipinnately compound

Leaf arrangement with leaflets along secondary petioles scattered along length of primary petiole.

725. blade

Broad, thin photosynthetic part of a leaf.

726. bract

Modified leaf associated with a plant's reproductive structures.

727. C-3 pathway

Carbon fixation in which first product is 3-carbon molecule.

728. C-4 pathway

Carbon fixation in which first product is 4-carbon molecule.

729. calorie

Quantity of heat needed to raise the temperature of one gram of water one degree Celsius.

730. chemolithotroph

Organisms that derive energy from inorganic compounds.

731. chemosynthesis

A proces by which large molecules are made from small ones using energy from other chemicals.

732. chemotroph

An organism that gets energy from oxidizing inorganic or organic matter.

733. chlorophyll

A green pigment capable of capturing light energy.

734. compound leaf

Leaf with blade divided into leaflets.

735. cone bract

A modified leaf associated with the axis of a cone.

736. cyclic photophosphorylation

The capture of energy in chloroplasts with the formation of ATP through the activity of the cell's cytochrome system.

737. dark reaction

A part of photosynthesis that can occur in either light or dark and that transfers energy from light reactions.

738. dichotomous venation

Branching of leaf veins into approximately equal parts without fusion of branches.

739. evaporation

Changing of a substance from liquid to gaseous form.

740. fleshy scale leaf

Modified leaf that stores food and water for a plant.

741. granum

A stack of thylakoids within a chloroplast.

742. guttation

The forcing of water from a plant's leaf tips.

743. kilocalorie

Heat required to raise the temperature of one kilogram of water one degree Celsius.

744. leaf gap

Break in vascular cylinder of stem above junction of leaf trace at a node.

745. leaf scar

Impression in a branch left after a leaf falls.

746. leaf sheath

Sheath-like petiole that encircles a portion of a stem as in a grass.

747. leaf trace

Vascular tissue branch that enters a leaf from a stem.

748. leaf vein

A vascular structure within a leaf.

749. leaflet

Unit of a compound leaf.

750. light reaction

Events in photosynthesis that capture energy and occur in light.

751. lobed leaf

Leaf with deep indentations in its blade.

752. megaphyll

A large broad leaf with greatly branched veins.

753. mesophyll

Photosynthetic tissue between the upper and lower surfaces of a leaf.

754. microphyll

A small leaf.

755. midrib

Large leaf vein that divides leaf into halves.

756. netted venation

Vein arrangement that forms a network in a leaf usually seen in dicotyledonous plants.

757. nodal bract

Modified leaflike structure projecting from a node on a cone or stem.

758. noncyclic photophosphorylation

The capture of energy in chloroplasts with the formation of ATP and reduced NADP.

759. opposite leaf arrangement

Distribution of leaves in pairs opposite each other at nodes.

760. palmately compound

Leaf having leaflets arranged in fan shape at tip of primary petiole.

761. palmately netted

Leaf vein arrangement with two or more large veins of equal size originating from base of leaf.

762. parallel venation

Leaf vein arrangement with several veins of equal size nearly parallel and longitudinal.

763. photoauxotroph

An organism that uses light energy to manufacture its own food.

764. photolithotroph

An organism that uses light energy to synthesize food from inorganic substances.

765. photophosphorylation

The addition of phosphate groups and high energy bonds to a molecule during the capture of light energy.

766. photosynthesis

The process by which organisms capture light energy from the environment and store it in a usable form.

767. photosystem

Chlorophyll molecules trapping light energy and tranferring it to chlorophyll participating in electron transfers.

768. pinnately compound

Leaflet arrangement along primary petiole in compound leaf.

769. pinnately netted

Leaf venation with large midrib vein and networks on either side.

770. pulvinus

Small enlargement at the base of a leaflet or petiole.

771. quantum

Smallest increment or parcel of many forms of energy.

772. rachis

Axis of an inflorescence or compound leaf.

773. sessile

Stalkless structure.

774. simple leaf

Leave with blade not divided into leaflets.

775. spine

Needlelike modified leaf.

776. stipule

Leaflike structure at the base of a leaf stalk in some plants, such as roses.

777. stoma

A pore in lower leaf epidermis through which gases diffuse into and out of mesophyll spaces.

778. stomatal chamber

Large air space associated with stoma.

779. sunken stomata

Stomata in which guard cells lie in a depressed area.

780. tenacious bract

Bract that clings to another bract.

781. thylakoid

Parallel flattened sacs that form part of the membrane structure of a chloroplast.

782. turbidity

Opaqueness or light-blocking capacity.

783. vein

Vascular tissue of a leaf.

784. venation

Arrangement of veins in a leaf.

785. whorl

Arrangement of three or more leaves at a nodal region.

Respiration and Cell Metabolism

786. **aerobic**

In the presence of oxygen.

787. **aerobic respiration**

Breakdown of organic materials using oxygen as electron acceptor.

788. **alcoholic fermentation**

Anaerobic metabolism of glucose to alcohol and carbon dioxide.

789. **amylase**

An enzyme that digests starch.

790. **anabolic**

Of anabolism.

791. **anabolism**

Synthetic, energy using process.

792. **anaerobic**

Lacking oxygen.

793. **anaerobic respiration**

Breakdown of organic molecules in absence of oxygen.

794. **catabolism**

Breakdown of molecules that makes energy available.

795. **cohesion**

The attraction of water molecules for each other.

796. deamination

Removal of an amino group.

797. deoxyribonuclease

An enzyme that digests DNA.

798. dormancy

State of decreased physiological activity.

799. electron transport system

Enzymes and coenzymes in cristae of mitochondria that move electrons from substrates to oxygen.

800. energy

The ability to do work.

801. enzyme

A protein that increases the rate of a chemical reaction in a living organism.

802. facultative anaerobe

Organism that normally uses free molecular oxygen but that can survive without it.

803. fermentation

An anaerobic metabolic process in which carbohydrate is broken down to alcohol and other simple molecules.

804. flavin adenine dinucleotide (FAD)

A coenzyme that carries hydrogen.

805. glycolysis

Metabolic pathway for breakdown of glucose to pyruvic acid.

806. hydrogen carrier

Molecule that can transport, donate, or accept hydrogen atoms or ions.

807. Krebs cycle

Metabolic pathway in which acetyl CoA is metabolized to carbon dioxide and water.

808. lipase

An enzyme that breaks down lipids.

809. maltase

Enzyme that digests maltose, a disaccharide derived from starch.

810. mesotrophic

Using ammonia or a particular amino acid as a nitrogen source.

811. metabolism

All chemical reactions in a living organism.

812. nicotinamide adenine dinucleotide (NAD)

A coenzyme that transports hydrogen atoms or electrons in oxidation-reduction reactions.

813. oxidative phosphorylation

Capture of energy in ATP during oxidative metabolism.

814. phosphorylation

Binding of a phosphate group to a molecule.

815. photorespiration

Oxygen use and carbon dioxide release stimulated by light without ATP production.

816. respiration

The processes of ventilation (breathing) and gas exchange.

817. ribonuclease

Enzyme that digests RNA.

818. sink

Region in a plant where food is used.

819. source

Area in a plant from which food can be translocated.

820. sucrase

An enzyme that digests sucrose.

821. transamination

Transfer of an amino group from one molecule to another.

822. turnover

Reuse of a substance made available by a catabolic reaction.

Growth, Development, Life Cycles

823. **abscisic acid**

A plant hormone that inhibits growth and induces dormancy in buds and seeds.

824. **abscission**

Separation and dropping of leaves, fruits, and other plant parts.

825. **afterripening**

Enzymatic action in seeds, fruits, tubers, and bulbs after harvest that allows later germination.

826. **air layering**

Vegetative propagation by placing a shoot or branch in moist soil or peat until roots form.

827. **alternation of generations**

A reproductive cycle involving gametophyte and sporophyte generations of plants.

828. **apogamy**

Development of an embryo without fusion of gametes.

829. **apogeotropic**

Root growth away from the earth and gravity.

830. **assay**

Procedure for determining quantity of a substance such as a hormone.

831. **autoecious**

Requiring a single host for completing a life cycle, as in wheat rust.

832. **auxin**

A plant hormone that stimulates cell division and cell growth.

833. cell plate

A structure between two plant cells at the end of mitosis where cell membrane and cell wall will develop.

834. centripetal

Developing from the surface toward the center.

835. chemotropism

The movement of cells or organisms toward or away from certain chemical substances.

836. circadian rhythm

Natural rhythm about a day in length.

837. critical daylength

Daylength required by a plant with photoperiodicity.

838. day neutral

Not subject to photoperiodicity.

839. determinate

Limited kind of growth pattern.

840. diaspore

Any plant part that can be dispersed.

841. differentiation

The specialization of structures during embyronic development.

842. diffuse growth

Elongation by cell division throughout a plant.

843. elicitor

A substance produced by a plant pathogen that induces the host plant to make resistant phytoalexins.

844. embryo

Developing multicellular structure derived from a zygote.

845. embryo sac

Female gametophyte of angiosperm with eight nuclei.

846. embryogeny

Formation of an embryo.

847. embryonic axis

Vertically elongated structure of a seed to which cotyledons are attached.

848. embryophyte

Plant in which the embyro develops where it arose.

849. ethylene

A simple organic molecule that stimulates fruit ripening.

850. evagination

Unsheathing; outgrowth.

851. geotropism

A plant's response to gravity.

852. gibberellin

A plant hormone that stimulates stem growth and affects flowering, root formation, and leaf growth in some plants.

853. growth ring

Secondary xylem produced during one growing season.

854. heteroecious

Requiring two hosts to complete a life cycle, as in some rust fungi.

855. hormone

Chemical substance from one kind of cell that influences growth and development of other cells.

856. leaf axil

The point at which a leaf and stem unite.

857. long-day plant

Plant requiring longer period of light than the critical day length to flower or make other physiological response.

858. morphogenesis

Development of form.

859. photoperiodism

Action of day length in regulating plant growth and development.

860. phototrophism

The bending of a plant toward light.

861. phytochrome

Blue pigment that controls some light-mediated growth and flower production in some plants.

862. plumule

The first shoot and leaves of a plant.

863. radicle

An embryonic structure in a seed that gives rise to a root.

864. secondary growth

Growth by cell division in lateral meristems.

865. seed

An embryonic plant consisting of the embryo, stored food, and a coat; a mature ovule.

866. senescence

Aging.

867. short-day plant

Plant requiring shorter light periods than critical day-length to produce a flower or other physiologic response.

868. substrate

Surface on which an organism grows.

869. target cells

Cells whose growth and development are controlled by a hormone.

870. thigmotropism

Growth in response to touch.

871. totipotent

Having the potential to give rise to any mature cell type.

872. unlimited growth

Indeterminate growth, which occurs indefinitely in the life of a plant.

Reproduction

873. antheridium

A male structure in some nonseed plants in which swimming sperm are produced.

874. apomixis

Reproduction with meiosis or gametes.

875. asexual

Reproduction not involving the union of gametes.

876. bisexual

Producing both male and female gametes.

877. budding

Asexual reproduction in which a new organism or cell pinches off from the parent.

878. cloning

Asexually produced progeny of a plant, naturally or by cuttings.

879. conjugation tube

Passage through which nuclei of mating cells pass.

880. diploid

Having paired chromosomes.

881. egg

Nonmotile female gamete.

882. entomophily

Pollination by an insect.

883. fertilization

Union of egg and sperm.

884. fission

Asexual reproduction in which a cell divides into two.

885. gametangium

A gamete producing structure in a plant.

886. gamete

Haploid cell; an ovum or sperm.

887. gametogenesis

The process of forming gametes.

888. gametophyte

The haploid, gamete producing portion of a plant life cycle.

889. gamone

A chemical substance that fosters union of gametes.

890. germination

Start of new growth by a reproductive structure.

891. gonad

An ovum or sperm.

892. hermaphrodite

Organism having organs of both sexes, or both male and female parts in the same flower.

893. heterogamy

Fusion of morphologically different gametes.

894. heteromorphic

Having a life cycle in which diploid and haploid phases differ in shape.

895. heterosporous

Producing different kinds of spores, some developing into female and others into male gametophytes.

896. heterospory

Having two functionally different spores produced by meiosis during the life cycle.

897. heterothallic

Production of male and female gametes by separate organisms.

898. homospory

Having a single type of spore produced by meiosis during the life cycle.

899. homothallic

Production of male and female gametes by the same organism.

900. isogamy

Having structurally identical male and female gametes.

901. mating type

A strain of sexually compatible organisms.

902. megagametophyte

Egg-producing part of a life cycle.

903. megasporangium

A sporangium that produces megaspores.

904. megasporophyll

A leaf that produces megaspores.

905. meiocyte

A diploid cell that undergoes meiosis; meiospore mother cell.

906. meiosis

Cell division that gives rise to haploid cells.

907. microgametophyte

Sperm-producing part of a life cycle.

908. micropyle

Opening in integument of an ovule through which pollen enters.

909. microsporangium

Sporangium in which microspores are produced.

910. microsporophyll

A microspore-producing leaf.

911. neck

Slender part of archegonium along which male gamete passes to reach the female gamete.

912. oocyte

Cell that gives rise to an ovum.

913. oogenesis

Process of producing an ovum.

914. oogonium

A mitotically dividing female germ cell that produces primary oocytes.

915. ornithophily

Pollination by a bird.

916. ovary

A female gonad.

917. ovum

Female gamete.

918. parthenocarpic

Concerning fruits produced without fertilization and that are seedless.

919. pollination

Transfer of pollen from anther to stigma in angiosperms or from microsporangium to ovule in gymnosperms.

920. pollinator

Agent that carries pollen from male to female parts.

921. progeny

Offspring.

922. propagation

New plant production sexually or asexually.

923. propagule

That which propagates, or produces, a new plant.

924. reductional division

First meiotic division in which chromosome numbers is reduced in half.

925. sexual

Reproduction involving the production and union of gametes.

926. soredium

Asexual reproductive organ of a lichen containing algal cells surrounded by fungal hyphae.

927. sperm

A male gamete.

928. spermatogonium

Undifferentiated male germ cells from which sperm-producing cells arise.

929. sporangiophore

A special branch bearing sporangia.

930. sporangiospore

Spore produced in a sporangium.

931. spore

A reproductive cell capable of producing a new individual without uniting with another cell.

932. spore mother cell

Diploid cell in which meiosis takes place.

933. sporic meiosis

The occurrence of meiosis during spore formation.

934. sporocarp

Structure containing spores; fruiting body in a fungus.

935. sporocyte

A cell that can produce one or more spores.

936. sporogenesis

The production of spores.

937. sporophore

Structure on which a sporangium develops.

938. sporophyll

A sporangium bearing leaf.

939. sporophyte

The diploid, spore producing generation in alternation of generations.

940. sporulation

Spore release.

941. strobilus

Conical structure comprised of leaves bearing sporangia.

942. synapsis

Alignment of homologous pairs of chromosomes.

943. syngamy

Fusion of gametes in fertilization.

944. tropism

A fixed kind of movement in response to a stimulus.

945. unisexual

Able to produce a single kind of gamete.

946. vegetative propagation

Asexual reproduction in flowering plants as from cuttings.

947. zoospore

A flagellated, motile reproductive cell involved in asexual reproduction in some algae and fungi.

948. zygospore

Thick-walled diploid spore capable of surviving adverse conditions.

949. zygote

Single cell resulting from a union of ovum and sperm; first cell of a new individual.

950. zygotic meiosis

The occurrence of meiosis during the division of a zygote in sexually reproducing organisms.

Taxonomy

951. Anthophyta

Division to which all flowering plants belong.

952. Arthrophyta

Division to which horsetails belong.

953. Ascomycetes

Class of fungi with ascus where meiosis and nuclear fusion occur.

954. Bryophyta

Division containing primitive land plants such as liverworts and mosses.

955. Chlorophyta

Division containing green algae.

956. Chrysophyta

Division containing golden-brown algae.

957. class

Taxonomic category within divisions that incorporates orders.

958. composite

Member of Compositae family; sunflower, artichoke, daisy.

959. Coniferophyta

Division containing conifers.

960. conspecific

Of same species.

961. cucurbitaceous

Member of Cucurbitaceae family; squash, melon, cucumber.

962. Cycadophyta

Division containing cycads.

963. Deuteromycetes

Class containing fungi of unknown sexual reproduction.

964. division

Taxonomic unit of plant classification comparable to phylum in animal taxonomy.

965. Euglenophyta

Division containing euglenoid algae.

966. family

Classification unit within an order and including several genera.

967. generic

Pertaining to a genus, as opposed to a particular species.

968. Ginkgophyta

Division containing ginkgos, a kind of gymnosperm.

969. Gnetophpyta

Division containing Ephedra, Gnetum, and certain other gymnosperms.

970. gymnosperm

Plant in which seeds develop in a cone.

971. lauraceous

Member of Lauraceae family; cinnamon, avocados, bay.

972. **legume**

Member of Leguminosae family; beans and peas.

973. **Microphyllophyta**

Division containing club mosses.

974. **Monera**

Kingdom containing all prokaryotic organisms.

975. **Myxomycota**

Division containing slime molds.

976. **nomenclature**

Naming of organisms.

977. **order**

Classification level composed of families.

978. **Phaeophyta**

Division containing all brown algae.

979. **Plantae**

Kingdom consisting of all true plants.

980. **Protista**

Kingdom containing all eukaryotic algae and all protozoa.

981. **Psilophyta**

Division containing whisk ferns.

982. **Pterophyta**

Division containing ferns.

983. Pyrrophyta

Division containing dinoflagellates.

984. Rhodophyta

Division containing red algae.

985. rosaceous

Member of Rosaceae family; roses, cherries, apples.

986. Schizophyta

Division containing bacteria.

987. sedge

Member of Cyperaceae family; monocotyledons with solid stems.

988. systematic botany

Branch of botany that deals with classification.

989. taxon

Any unit of classification.

990. Zygomycetes

Class of fungi with nonseptate hyphae and spores without flagella.

Evolution

991. abiogenesis

Beginning without life; spontaneous generation.

992. adaptive radiation

Change in a population over time such as by divergence or convergence.

993. allopatric speciation

The rise of new species from populations of a single species that were physically isolated from one another.

994. amber

Fossilized resin.

995. analogous structure

One of two or more structures having similar functions.

996. analogy

Similarity of organisms due to common form or function but not resulting from common ancestry.

997. artificial taxon

Taxonomic group of organisms related as closely to members of other groups as to members of the group itself.

998. binomial system

The two-name system of naming organisms.

999. biogenesis

Generating life from life.

1000. biological evolution

Changes over time in living organisms.

1001. cast

Fossil form preserved in a mold at the site where the organism decomposed.

1002. chemical evolution

The gradual increase in complexity of molecules thought to have preceded the origin of living cells.

1003. cladistics

Development of a classification based on shared derived traits.

1004. coacervate droplet

A mixture of large molecules thought to have preceded the organization of the first cells.

1005. common ancestor

An ancestor to two or more branches in the evolutionary tree.

1006. competitive exclusion principle

The principle that if two species continue to compete for the exact same resources, one will become extinct.

1007. continental drift

Movement of continents as ocean floor spreads.

1008. dichotomous key

A means of identifying organisms by chosing which of paired statements in a series pertain to an organism.

1009. differential migration

Alteration of gene frequencies in a population by population mobility.

1010. directional selection

Selection involving changes that occur when a population displays a steady trend over time.

1011. disruptive selection

Selection due to unusual features that have high survival value.

1012. divergence

Radiating out in different directions.

1013. ecological equivalent

An unrelated organism having functions similar to another in the same environment.

1014. evolution

The process of change over time.

1015. evolutionary tree

A diagram showing evolutionary relationships among selected organisms.

1016. extant

Living, as opposed to extinct.

1017. extinct

No longer having living representatives.

1018. five kingdom system

A taxonomic system that places all living organisms in one of five kingdoms.

1019. fossil

Any evidence of organisms that lived in the past.

1020. gene flow

The movement of genes from one population to another by reproduction between members of the populations.

1021. gene pool

The sum of all genes and their alleles present in a population at a given time.

1022. genetic drift

Fluctuations in gene frequencies due to isolation of nonrepresentative sampling of the founding population.

1023. genetic equilibrium

A state of constancy in the frequency of alleles in a population.

1024. genus

The taxonomic category that combines similar species; the first part of the scientific name of an organism.

1025. Hardy-Weinberg equilibrium

A state in which gene frequencies remain unchanged from generation to generation in a population.

1026. homology

Similarity in form and function because of common ancestry.

1027. homoplasy

Appearance of a similar feature in distantly related taxonomic groups.

1028. isolating mechanism

A factor that prevents matings between members of two populations.

1029. macroevolution

Large scale evolutionary change.

1030. microfossil

A particle such as a fossilized spore, pollen grain, or other plant part.

1031. microsphere

A structure consisting of protein but having certain attributes of a cell.

1032. monophyletic

Evolving from a single ancestral stock.

1033. natural taxon

Taxonomic group of organisms more closely related to each other than to organisms in any other group.

1034. numerical taxonomy

Development of classification schemes by numerical analysis of similarity of characteristics.

1035. ontogeny

Developmental stages of organisms of a species.

1036. oxidizing atmosphere

An atmosphere containing oxygen and other oxidizing molecules.

1037. parallel evolution

Similar adaptations to an environment in the evolution of unrelated organisms.

1038. phylogenetic tree

A branching diagram showing evolutionary relationships among selected organisms.

1039. phylogeny

Evolutionary history of a group of organisms.

1040. polymorphic

Pertaining to a species having two or more distinct kinds of individuals.

1041. population

A set of all interacting, interbreeding individuals of one species.

1042. primitive

Plant or plant characteristic similar to oldest know plant characteristics.

1043. primitive atmosphere

Atmosphere having gases that were present prior to the emergence of life on earth.

1044. primordial

Original or primitive.

1045. progenitor

Early ancestor in an evolutionary line.

1046. protocell

Precursor cell hypothesized to have arisen in primeval times.

1047. reproductive isolation

Separated by a physical or biological barrier that prevents sexual reproduction.

1048. scientific name

The genus and species to which an organism belongs.

1049. selection pressure

Competition and environmental factors that influence the survival and reproduction of an organism.

1050. speciation

The creation of a new species.

1051. species

A group of similar organisms having common genes.

1052. sympatric speciation

Origin of a new species from a group of an existing species because of physiological or behavioral isolation.

1053. taxonomy

The science of classifying organisms.

1054. vestigial structure

The remains of a structure that was once functional in an ancestor.

Prokaryotes

1055. actinoycete

Filamentous soil bacteria.

1056. akinete

Thick walled, nonmotile spore from some cyanobacteria.

1057. anisogamy

Having gametes of different sizes.

1058. antibiotic

Chemical substances produced by fungi that kill other organisms.

1059. archaebacterium

A primitive kind of bacterium.

1060. bacillus

A rod-shaped bacterium.

1061. bactericidal agent

An agent that kills bacteria.

1062. bacteriostatic agent

An agent that inhibits bacterial growth.

1063. bacterium

Prokaryotic organism typically nonphotosynthetic.

1064. binary fission

Asexual reproduction in which a cell or an organism separates into two cells.

1065. blue-green algae

Cyanobacteria; photosynthetic prokaryotes.

1066. coccus

A spherical bacterium.

1067. colony

An organism consisting of a loose collection of cells with a modest degree of specialization.

1068. conjugation

A sexual union in which nuclear material of one cell enters another.

1069. cyanobacteria

Primitive photosynthetic prokaryotic organisms; blue-green algae.

1070. cytostome

A permanent site in a ciliate through which food enters.

1071. diplococcus

A bacterium found in pairs of spherical cells.

1072. endospore

A spore formed by certain bacteria that allow them to survive unfavorable conditions.

1073. gram-negative

Bacteria that do not take up crystal violet stain.

1074. gram-positive

Bacteria that do take up crystal violet stain.

1075. heterocyst

A cell that can fix nitrogen and serve as an asexual spore found in some cynaobacteria.

1076. inoculum

A small sample of a culture of microorganisms used to start a new culture.

1077. nuclear body

Region containing DNA in a prokaryotic organism.

1078. pathogen

An organism that can cause disease in another organism.

1079. phage

Virus that infects a bacterium.

1080. phycobilin

A water soluble pigment found in cyanobacteria and red algae and related to bile pigments in animals.

1081. phycobilisome

A structure on a photosynthetic membrane that contains phycobilins.

1082. pilus

Small extensions from bacteria used in conjugation.

1083. retrovirus

A virus that contains RNA and uses it to make DNA.

1084. rickettsia

Microorganisms smaller than bacteria that reproduce only in cells.

1085. spirillum

Bent rod or spiral shaped bacterium.

1086. vibrio

A comma-shaped bacterium.

Fungi and Slime Molds

1087. abstrict

Discharge a plant part, such as spores from a basidiomycete.

1088. aecium

Cup-like structure in which aeciospores form.

1089. aggregation

Amoebic movement in a slime mold toward a single point.

1090. amoeboid

Continually changing shape and moving.

1091. arbuscle

Branched fungal structure in a root cell that exchanges nutrients.

1092. arthrospore

Spore formed by hyphal fragmentation.

1093. ascocarp

Fruiting body of ascomycete fungus.

1094. ascogonium

Female gametangium of ascomycete fungus.

1095. ascospore

A spore from a single-celled fungus capable of surviving long periods of drought and extreme temperature.

1096. ascus

A sac-shaped spore forming cell associated with sexual reproduction in certain fungi.

1097. basidiocarp

Fruiting body of basidiomycete fungus.

1098. Basidiomycetes

Class of fungi with basidium where meiosis and nuclear fusion occur.

1099. basidiospore

A spore produced by basidia in gills of mushrooms.

1100. basidium

A club-shaped structure in which spores are produced.

1101. bread mold

A fungus commonly found growing on bread.

1102. coenocyte

A multinucleate organism or one made of multinucleated cells.

1103. conidium

A spore produced by a sac or club fungus during asexual reproduction.

1104. dermatophyte

A fungus that grows in keratinized tissue (skin, hair, and nails).

1105. dikaryotic

Having two nuclei.

1106. fairy ring

A circle of mushrooms formed by radially growing mycelia.

1107. filament

The portion of a stamen that supports an anther.

1108. fruiting body

Reproductive structure that produces spores by meiosis in ascomycete or basomycete fungi.

1109. fungus

An organism of the kingdom containing molds and plantlike organisms lacking chlorophyll and feeding on detritus.

1110. gill

Mycelia on the underside of a basidiocarp.

1111. haustorium

A fungal hypha or other plant part that takes up nutrients.

1112. heterokaryotic

Having two distinct kinds of nuclei in same cytoplasm, such as haploid and diploid nuclei of some mycelia.

1113. homokaryotic

Having nuclei of the same genetic makeup.

1114. hypha

A threadlike extension of a fungus.

1115. hypothallus

Membranous film at the base of a fruiting body in a slime mold.

1116. imperfect stage

Fungal stage in which structures for asexual reproduction are produced.

1117. karyogamy

Conjugation of cells and union of nuclei.

1118. lichen

A symbiotic combination of an alga and a fungus.

1119. mantle

Mass of hyphae around a root.

1120. mycelium

A mass of hyphae making up the body of a fungus.

1121. mycorrhiza

Symbiotic relationship between fungi and plant roots.

1122. oogamy

Having gametes differentated as egg and sperm.

1123. perfect stage

Sexual stage of a fungus.

1124. peridium

Outer layer of sporangium of myxomycetes and some other fungi.

1125. plasmodium

A slimy mass of acellular cytoplasm.

1126. plasmogamy

Cytoplasmic fusion of cells.

1127. primary plasmodium

Plasmodium developing from a haploid swarm cell.

1128. pseudoplasmodium

Cell mass formed by aggregation of myxamoeba in cellular slime molds.

1129. pseudopod

An extension from the main cell mass of an amoeba.

1130. sporangium

A spore bearing structure on the underside of a fern leaf.

1131. swarm cell

Flagellated cell produced by spore germination in myxomycetes.

1132. swarming

Release of swarmers and their subsequent movements.

1133. yeast

Single-cell fungus of class Ascomycetes.

Algae

1134. agar

A complex carbohydrate derived from red algae and used to solidify bacterial growth media.

1135. alga

Simple photosynthetic organism lacking vascular tissue and having unicellular gametangia.

1136. algal bloom

A population explosion among algae, usually on a pond surface.

1137. allophycocyanin

Blue pigment in blue-gree algae.

1138. aplanospore

Zoospore of an alga that lacks flagella or nonmotile spore of a fungus.

1139. coenobium

Algal colony having a fixed number and arrangement of cells at maturity.

1140. contractile vacuole

Small pulsating organelle that regulates fluid balance in some algae and fungi.

1141. diatom

A unicellular alga with cell walls containing silica.

1142. diatomaceous earth

Deposit of mineral remains of diatoms.

1143. diplobiontic

Algae that produce two independent multicellular forms in their life cycle.

1144. ejectosome

Organelle that can be ejected by some flagellated algae.

1145. fucoidan

Phycocolloid found in cell walls and spaces in brown algae.

1146. fucosterol

A sterol found in some brown and red algae.

1147. fucoxanthin

A xanthophyll pigment in brown and golden-brown algae.

1148. fucoxanthin

Xanthophyll pigment found in some brown algae and chrysophytes.

1149. groove

Longitudinal depression in some algae.

1150. haplobiontic

Algal species that have only free-living forms in their life cycles.

1151. holdfast

Rootlike structure that attaches some algae to their substrate.

1152. kelp

Common name for a group of brown algae.

1153. phycobiont

Algal component of a lichen.

1154. phycocyanin

Blue, water-soluble photosynthetic pigment in red algae and cyanobacteria.

1155. phycoerythrin

A red photosynthetic pigment in red algae.

1156. red tide

Deposition of red neurotoxin by dinoflagellate that kills large numbers of fish.

1157. rhizine

A bundle of hyphae attaching a lichen thallus to its substrate.

1158. rockweed

Common name for Fucus, a kind of brown algae.

1159. theca

A case.

1160. trichogyne

Tube from female gametangia of red algae and ascomycetous fungi through which sperm pass.

Non-Flowering Land Plants

1161. amphithecium

Outer cells of embyronic sporangium of a bryophyte.

1162. annulus

Row of unevenly thickened cells around sporangium of a fern.

1163. antheridiophore

Stalk bearing antheridia.

1164. archegoniophore

Stalk to which archegonium is attached.

1165. archegonium

A female structure in certain nonseed plants in which an egg
is produced.

1166. bisporangiate cone

Cone containing both megaspores and microspores.

1167. bryology

Study of liverworts and mosses.

1168. bryophyte

A nonvascular plant such as a liverwort or moss.

1169. caulid

Main shoot of gametophore in a bryophyte.

1170. club moss

Primitive land plant of the Microphyllophyta.

1171. cone

Branch with sporophylls or ovuliferous scales attached; a strobilus.

1172. conifer

Major group of gymnosperms, such as pines and spruces.

1173. crozier

Young fern frond in process of unfolding.

1174. cupule

Cup-shaped structure holding ovules in pteridosperms.

1175. cycad

A cone-bearing palmlike plant found in the tropics.

1176. endothecium

Inner cells in early embryonic sporangium of a bryophyte.

1177. fern

Primitive vascular plant that does not produce seeds.

1178. free-sporing

Production of spores that are shed before they germinate.

1179. frond

Fern leaf.

1180. gemma

An asexually produced miniature of an organism, typical of bryophytes.

1181. gemma cup

Cavity in which gemma form.

1182. ground pine

Club mosses and other plants of the division
Microphyllophyta.

1183. gymnosperm

One of a group of vascular plants having naked or unenclosed
seeds that develop in a cone.

1184. horsetail

Primitive vascular plant having a strobilus and jointed
stems.

1185. hydroids

Cells specialized for conducting water in bryophytes.

1186. hydrome

Tissue that conducts water and minerals in bryophytes.

1187. indusium

Cover of a sorus on a fern leaf.

1188. innovation

Branch arising from an old stem in a bryophyte.

1189. leptoma

Thin area in gymnosperm pollen wall through which pollen
tube emerges.

1190. liverwort

Flat, lobed, succulent, nonvascular plant that grows near
soil; a bryophyte.

1191. moss

Small bryophyte with leafy appearance.

1192. nonvascular plant

Primitive plant lacking xylem and phloem tissues.

1193. peat

Partially decomposed deposits of dead mosses.

1194. perigonial branch

A branch that bears antheridia in a bryophyte.

1195. peristome

Ring of triangular cells surrounding the opening of a sporangium in a moss.

1196. phyllid

Flattened leaflike structure in a bryophyte.

1197. prothallial cell

Cell of a gymnosperm pollen grain of unknown function.

1198. prothallus

Autotrophic, multicellular gametophyte phase in the life cycle of ferns and other primitive vascular plants.

1199. protonema

A body of branching filaments in a spore.

1200. ramentum

Sterile, hair-like scales on seeds in some cones.

1201. resin duct

Tubular structure, mainly in gymnosperms, that transports resin.

1202. rhizoid

A root-like organ in moss plants.

1203. seta

Stalk of sporophyte generation of a bryophyte.

1204. sorus

A cluster of sporangia.

1205. stomium

Opening in sporangium or anther through which dehiscence occurs.

1206. tracheophyte

A plant with vascular tissue.

1207. vascular plant

A plant having xylem and phloem tissues.

1208. whisk fern

Member of the division Psilophyta.

Flowering Plants

1209. adventitious

Arising from other than the normal site.

1210. angiosperm

A flowering vascular plant with seeds in carpels that develop into fruits.

1211. ascidium

A cup-shaped organ in a flowering plant.

1212. axile placentation

Placenta arranged in center of compound pistil and having two or more chambers.

1213. basal placentation

Placenta arranged at lower end of carpel in simple pistil and with one ovule.

1214. betacyanin

Water-soluble red pigment found in cell vacuoles of a few flowering plants.

1215. bulb

A compressed stem bearing fleshy leaves.

1216. coleoptile

A sheath that covers the first leaves of a grass seedling.

1217. corm

A short, bulky stem containing stored food.

1218. cotyledon

A seed leaf in a plant embryo.

1219. cross-pollination

Pollination involving the transfer of pollen from one plant to another.

1220. determinate umbel

Umbel having branches that do not subdivide.

1221. dicotyledon

A plant whose embryo had two seed leaves.

1222. dioecious

Having staminate and pistillate flowers on separte plants.

1223. endosperm

A nutrient material in a plant embyro.

1224. endosperm nucleus

Nucleus that will give rise to endosperm.

1225. epicotyl

Tiny leaves and a bud found inside a typical dicot seed.

1226. epigynous

Above the gynoecium in an angiosperm.

1227. flavone

Group of water-souble plant pigments that give flowers their colors.

1228. floral apex

Apical meristem that will give rise to a flower or inflorescence.

1229. flower

Aggregation of spore-bearing and sterile parts in a flowering plant.

1230. free central placentation

Placenta arranged around free column of tissue fixed to the base of a compound ovary.

1231. free-nuclear endosperm

Endosperm made from multiple nuclear divisions without cytokinesis and before formation of a cell wall.

1232. grain

Dry fruit of grasses in which seed is fused with fruit wall.

1233. heartwood

Dark inner xylem of a woody root or stem that is unable to transport water and minerals.

1234. hypocotyl

A minature embryo stem found inside a typical dicot seed.

1235. hypogynous

Below the gynoecium in an angiosperm.

1236. hypsophyll

Structure consisting of modified leaves associated with a flower.

1237. incomplete flower

A flower lacking one or more parts that make up a complete flower.

1238. indeterminate

An inflorescence having youngest flowers at the top.

1239. inflorescence

Group of flowers on a common stalk.

1240. marginal placentation

Placenta to which ovules are attached lies along line where two margins of carpels meet.

1241. monocotyledon

A kind of angiosperm whose embryo has only one seed leaf.

1242. monoecious

Having separate staminate and pistillate flowers on the same plant.

1243. nectary

Organ that secretes nectar.

1244. nut

Dry, indehiscent, one-seeded hard fruit.

1245. parietal placentation

Placenta arranged along margins where adjacent carpels fuse.

1246. parthenocarpy

Production of seedless fruit without fertilization.

1247. placenta

Tissue from which ovules are produced in an ovary.

1248. placentation

Arrangement of placenta in an ovary.

1249. primary endosperm cell

Cell resulting from fusion of sperm with two polar nuclei to make endosperm.

1250. receptacle

End of a flower stalk where floral parts develop.

1251. scarification

Mechanical abrasion of seed coat.

1252. self-pollination

Transfer of pollen from the anther to the stigma of same flower or flowers on same plant.

Flowers

1253. achene

Dry, single-seeded, indehiscent fruit.

1254. aggregate fruit

Fruit derived from several pistils of same flower.

1255. albuminous

Describing seeds that contain endosperm at maturity.

1256. aleurone layer

Layer of the seed of a cereal grain.

1257. androecium

Collectively, all male reproductive structures in a flower.

1258. anther

The part of a stamen in which pollen is formed.

1259. anthesis

Flowering time.

1260. antipodal

One of nuclei in angiosperm megagametophyte located at end opposite micropyle.

1261. aril

Fleshy seed covering.

1262. berry

Simple fruit in which exocarp forms skin that covers fleshy mesocarp and endocarp.

1263. calyx

Collectively, all the sepals of a flower.

1264. capsule

Fruit from drying after fusion of two or more carpels.

1265. carpel

A simple pistil or unit of a compound pistil.

1266. catkin

Inflorescence with unisexual flowers arranged on a stalk.

1267. chalaza

Part of ovule next to stalk and opposite to micropyle.

1268. complete flower

A flower with sepals, petals, stamens, and a pistil.

1269. compound pistil

Pistil made by fusion of two or more carpels.

1270. corolla

The entire whorl of a flower's petals.

1271. corolla tube

Tube derived from fusion of the petals of a flower.

1272. corymb

Inflorescence of flower stalks of unequal length on the peduncle producing a flat-top effect.

1273. cyme

Inflorescence with younger flowers further from the apex which has terminal flower.

1274. dehiscence

Manner of opening at maturity.

1275. dehiscence zone

Area in a fruit where fruit wall ruptures and releases seeds.

1276. dehiscent

Splitting open at maturity.

1277. double fertilization

The two fertilizations that form the zygote and endosperm in flowering plants.

1278. drupe

Fleshy fruit with stony seed.

1279. egg apparatus

Ovum and two synergid nuclei at the micropylar end of a megagametophyte.

1280. fruit

A mature ovary of a flowering plant.

1281. generative cell

The cell in a pollen grain that generates a sperm.

1282. gynoecium

Collectively, the pistils in a flower.

1283. head

Inflorescence with a cluster of sessile flowers on a single receptacle.

1284. helicoid cyme

A determinate inflorescence with all lateral branches on the same side.

1285. imperfect flower

Flower having structures of a single sex.

1286. indehiscent fruit

Dry fruit that does not split open even when mature.

1287. indeterminate head

Flat inflorescence having youngest flowers in the middle.

1288. infrutescence

Fruit-bearing structure.

1289. integument

Outer cover of the nucellus of an ovule, which develops into a seed coat.

1290. irregular flower

Flower with flower parts asymmetrically arranged.

1291. megaspore

Haploid cells, one of which survives to form a female gametophyte In higher plants.

1292. megaspore mother cell

Diploid cell that gives rise to four haploid megaspores.

1293. mesocarp

Middle layer of pericarp.

1294. microspore

A spore that develops into a pollen grain in seed plants.

1295. microspore mother cell

A cell in the anther that produces four microspores.

1296. multiple fruit

Fruit composed of ovaries of many flowers.

1297. nucellus

Tissue around the megagametophyte in the ovule of a seed plant.

1298. outer integument

Outermost of two layers of coverings on an angiosperm ovule.

1299. ovule

The part of a plant ovary that develops into a seed.

1300. panicle

Inflorescence with flowers arranged alternately on branched stalk.

1301. pappus

A scale-like calyx.

1302. pedicle

Stalk of each flower in an inflorescence.

1303. peduncle

Stalk of a solitary flower or an inflorescence.

1304. perfect flower

Flower having parts of both sexes.

1305. perianth

External cover of a flower; calyx and corolla.

1306. pericarp

Wall of a fruit.

1307. petal

A part of the corolla, usually a showy part of a flower.

1308. pistil

The ovary, style, and stigma found together in the center of a flower.

1309. polar nucleus

One of two central nuclei in female gametophyte of angiosperm that fuse with a sperm to become endosperm nucleus.

1310. pollen chamber

Flask-like chamber at the top of the nucellus of an ovule.

1311. pollen grain

An immature male gametophyte consisting of a generative nucleus and a tube nucleus.

1312. pollen tube

Extension of a tube cell as pollen grain germinates that transports sperm to ovule.

1313. raceme

Inflorescence with flowers arranged alternately on unbranched stalk.

1314. regular flower

Flower with parts symmetrically arranged.

1315. scorpioid cyme

Determinate inflorescence with alternating lateral flowers and oldest flowers at base.

1316. secondary nucleus

Nucleus made by fusion of polar nuclei before it fuses with sperm nucleus.

1317. sepal

The outer protective part of a flower.

1318. simple fruit

Fruit derived from simple or compound pistil of a single flower.

1319. simple pistil

Pistil with one carpel.

1320. spike

Inflorescence with sessile flowers arranged alternately on an unbranched stalk.

1321. sporopollenin

Complex, sturdy substance comprising outer walls of pollen grains.

1322. stamen

The male part of a flower, including filament and anther.

1323. stamen tube

Tube produced by fusion of stamens.

1324. staminate flower

Flower that has a stamen and lacks pistils.

1325. stigma

A sticky structure in the pistil of a flower that receives pollen grains.

1326. style

A part of the pistil in a flower that supports the stigma.

1327. superior ovary

Ovary with relatively higher position with with respect to other flower parts.

1328. testa

Seed coat.

1329. thyrse

Inflorescence with determinate lateral axes and indeterminate main axis.

1330. tube cell

Cell of pollen grain that produces pollen tube.

1331. umbel

Inflorescence with all flowers at same height and all stalks originating from same point.

Weather, Climate, Soils

1332. alluvial

Concerning soil deposited by running water; silt, clay, and sand.

1333. anticlone

High pressure area in atmosphere.

1334. climate

Prevailing weather conditions in a region.

1335. cloud seeding

Dropping of substance such as dry-ice onto a cloud to foster rain.

1336. convection

Circulatory motion in gas or fluid due to effects of temperature, density, and gravity.

1337. convective current

Mass movement of air, with hot air rising and cool air falling.

1338. coriolis force

Curvature of winds to right in northern hemisphere and to left in southern hemisphere, due to earth's rotation.

1339. ebb tide

Period between high water and low water in tidal cycle.

1340. edaphic

Concerning drainage and soil conditions in which a plant is found.

1341. flood tide

Period of increasingly higher tides caused by the moon's orbit.

1342. frost line

Depth into the ground to which frost penetrates.

1343. hummock

Small mound of ground above a generally level terrain.

1344. infrared

Wavelengths longer than visible light felt as heat.

1345. laterite

Clay soil, which has been leached at high temperature.

1346. orographic

Concerning mountains.

1347. percolation

Filtration through porous material, such a soil or gravel.

1348. perlite

Soil substitute derived from volcanic rock.

1349. physiography

Description of the earth's surface.

1350. plant food

Minerals and other substances plants need for growth.

1351. podzolic soil

Soil containing a thin acidic layer and an organic mat.

1352. soil structure

Nature and arrangement of particles in soil.

1353. soil texture

Relative numbers of different sizes of particles in soil.

1354. solstice

Time of year when the sun is farthest from the equator; to the north in summer and to the south in winter.

1355. stratosphere

Region of atmosphere containing the ozone layer.

1356. thermocline

A sharp temperature difference between layers in a body of water that prevents distribution of oxygen and nutrients.

1357. topography

Surface configuration.

1358. trace element

Mineral or other element needed in small quantities.

1359. trophosphere

Band 7 to 10 miles wide below stratosphere in which most weather events occur.

1360. vermiculite

Soil substitute derived from the mineral mica.

1361. weather

Atmospheric condition with respect to wind, temperature, and moisture.

Concepts of Ecology

1362. allelopathy

Detrimental effect of one plant's secreted toxins on neighboring plants.

1363. association

A fundamental unit in an ecological community.

1364. biological oxygen demand

Amount of oxygen needed by organisms in a body of water.

1365. biomass

The total mass of all organisms living in a particular location.

1366. carbon cycle

A repetitive sequence of chemical processes in which carbon enters and leaves living organisms.

1367. carnivorous

Eating animal flesh.

1368. carrying capacity

The number of individuals of a species a particular environment can support indefinitely.

1369. climax community

Organisms that maintain ecological balance that favors their continued survival; final community in succession.

1370. commensalism

The relationship of two species in which one benefits and the other is neither benefitted nor harmed.

1371. consumer

An organism of one population that feeds on organisms of other populations within an ecosystem.

1372. decomposer

An organism that feeds on the remains of other organisms within an ecosystem.

1373. denitrification

The process of converting nitrate to nitrogen gas; a part of the nitrogen cycle.

1374. density-dependent factor

A population control factor that has a greater effect as the population size increases.

1375. density-independent factor

A population control factor that has the same effect regardless of the size of the population.

1376. detritus

Nonliving organic matter.

1377. ecological niche

The position of an organism in its environment; its diet, predators, habitat, and effects on its environment.

1378. endosymbiont

An organism that lives with another in a symbiotic association.

1379. environmental resistance

The carrying capacity of an environment for a species, stated as environment's resistance to population growth.

1380. erosion

Soil loss due to wind or rain action.

1381. etiolization

Yellowish, spindly stems and poorly developed leaves in plants grown in the dark.

1382. eutrophication

The process of aging and death of organisms in a pond or lake.

1383. exponential growth

Growth that frequently doubles the population size.

1384. food chain

The flow of energy and matter from the environment through organisms within an environment.

1385. food pyramid

A way of depicting the dependence of animals on other organisms in an ecosystem.

1386. food web

A pattern of interconnected food chains.

1387. fossil fuel

Combustible material, such as coal, oil, gas, derived from previously living material.

1388. geometric growth

Increase in population by a constant percentage in each generation.

1389. groundwater

Water in soil and in aquifers beneath the soil.

1390. habitat

The place where an organism normally lives.

1391. herbivore

An animal whose diet consists mainly of plants.

1392. heterotroph

An organism that metabolizes ready made organic matter.

1393. home range

The region to which an animal or a small group of animals normally confines its activities.

1394. host

Organism that provides for the needs of a parasite.

1395. humus

Partially decomposed remains of organisms and their wastes.

1396. hydrothermal vent

A volcanic opening on the ocean floor from separation of tectonic plates that spews forth hot water and minerals.

1397. irruptive growth

Population growth characterized by exponential growth, followed by catastrophic population reductions.

1398. mutualism

A relationship between two organisms of different species that benefits both organisms.

1399. net productivity

Amount of energy stored per square meter of land surface.

1400. niche

The role an organism plays in its ecosystem.

1401. nitrification

Formation of nitrates by bacteria.

1402. nitrogen cycle

A sequence of repeated reactions that move nitrogen between organisms and their environment.

1403. nitrogen fixation

A process that chemically combines atmospheric nitrogen with other elements.

1404. obligate parasite

Organism that must obtain what it needs from another organism.

1405. overturn

Mixing of a body of water caused by changes in temperature of layers and other external factors.

1406. parasite

An organism that resides in or on another organism and does harm to the host organism.

1407. parasitism

A symbotic relationship in which one organism lives at the expense of and does some damage to the host organism.

1408. phytoplankton

Small and microscopic aquatic (floating/motile) autotrophs.

1409. pioneer species

The first organisms to become established in an area at the beginning of ecological succession.

1410. plankton

A collection of free-floating, aquatic organisms carried by fresh water or ocean currents.

1411. plant community

Group of plants growing in a particular area.

1412. plant ecology

Study of interactions of plants and other organisms in the environment.

1413. population density

The number of one kind of organism in a defined area.

1414. predation

The eating of one organism by another.

1415. prey

An organism eaten by a predator.

1416. primary consumer

An organism that consumes plant material.

1417. primary producer

Photosynthetic organism that uses light energy and inorganic materials to make organic molecules.

1418. producer

An organism that can make organic nutrients from inorganic substances in the environment.

1419. replacement reproduction

A reproductive rate that replaces individuals that die and maintains a constant population size.

1420. secondary consumer

An animal that eats animals that have fed on plants.

1421. seral stage

Any of the stages of ecological succession in an ecosystem.

1422. sere

Any stage in ecological succession of an ecosystem.

1423. siltation

Deposition of silt.

1424. stratification

The layering of subcommunities as in a soil layers.

1425. succession

A series of ecological stages by which the community in a particular area gradually changes.

1426. symbiosis

A relationship between two interacting species of organisms.

1427. tolerance

Ability to resist effects of toxic substances.

1428. trophic level

A categorization of species in a food web according to how they obtain nutrients.

1429. water cycle

A sequence of repetitive reactions in which water moves to and from living things.

1430. water table

A level below which aquifers are filled with water.

1431. zonation

Divisions of a natural community according to variations in physical conditions.

Ecosystems and Biomes

1432. aphotic zone

Region without light deep in a body of water.

1433. benthos

Region at the bottom of an aquatic habitat.

1434. biome

A major area of the earth having certain life forms maintained by the climate of the region.

1435. bog

Poorly drained, acidic region rich in remains of dead organisms.

1436. boreal

Concerning northern biomes, such as tundra.

1437. brackish

Having salinity lower than in the ocean.

1438. canopy

Arrangement of top branches in the trees of a forest.

1439. chapparal

A biome containing broad leaved evergreen shrubs in a dense thicket.

1440. coral reef

Barrier made by secretions of skeleton of coral polyps.

1441. crown

Highest foliage in shrub or tree.

1442. desert

An area receiving 25 cm or less of rain per year.

1443. desertification

The process by which an area of marginally useful farm land becomes a desert by overgrazing or other abuses.

1444. diversified habitat

Region having a large variety of plant species.

1445. ecotone

Transitional region between two ecological communities.

1446. ecotype

Population adapted to a local habitat in which adaptations are transferred genetically.

1447. emergent

Extending above water level or forest canopy.

1448. epilimnion

Water layer overlying the thermocline of a lake and that is disturb by wind action.

1449. estuary

A body of water containing a mixture of fresh and salt water.

1450. eulittoral

Landward region of the littoral zone of a body of water.

1451. euphotic zone

The surface layer of a body of water to the depth penetrated by light; region in which photosynthesis occurs.

1452. fen

Low, peaty land partially or completely covered with water.

1453. grassland

A biome occupied mainly by grasses and having a dry climate.

1454. hypolimnion

Water layer beneath the thermocline, which is stagnant except during an overturn.

1455. intertidal region

Part of continental shelf exposed between high and low tides.

1456. lentic

Pertaining to a lake or pond.

1457. littoral zone

Region on or near shore, especially in ocean.

1458. lotic

Pertaining to a creek or river (running water) environment.

1459. marine

Pertaining to the ocean.

1460. oligotrophic

Pertaining to an environment having little nutrients available to organisms.

1461. ongotrophic

Concerning a lake that lacks plant nutrients.

1462. pelagic zone

Open ocean.

1463. permafrost

Perennial frozen subsoil in Arctic or subarctic regions.

1464. photic zone

Surface layer of a body of water in which photosynthesis can occur.

1465. phytotoxicity

Quality of being poisonous to plants.

1466. pocosino

Marsh or swamp, usually on a coastal plain.

1467. prairie

A temperate grassland biome having rainfall between 25 and 40 cm per year.

1468. rain forest

Evergreen forest found in regions with more than 100 inches of rain per year.

1469. riparian

On the bank or shore.

1470. salt marsh

Coastal grassland that undergoes seasonal flooding.

1471. savanna

A grassland biome that has occasional trees especially in Africa.

1472. secondary succession

Plant growth in a region partially cleared by fire or human action.

1473. steppe

Arid region with extreme temperature variation.

1474. sublittoral

On the water side of a shoreline; between rooted vegetation and the hypolimnion.

1475. submergent

Completely or partly covered by water.

1476. subtidal region

Part of continental shelf always covered by water.

1477. taiga

A northern forest containing mainly conifer trees.

1478. temperate deciduous forest

A biome with a moderate climate and many large trees that lose their leaves.

1479. timberline

Altitude beyond which trees fail to grow because of harsh climate.

1480. tropical rain forest

A hot moist environment with a very large number of species.

1481. tundra

A cold biome with permafrost and lacking trees.

1482. understory

Foliage layer beneath canopy.

1483. wetland

Swamp or march that has standing water most of the year.

Pollution

1484. acid rain

Rain containing acids from atmospheric pollution.

1485. activated sludge

Product of secondary sewage treatment treated with air so aerobes can break it down.

1486. biodegradable

Capable of being decomposed by living organisms.

1487. biological control

The use of one organism to control another, especially the reduction in numbers of an undesirable organism.

1488. biological magnification

The concentration of substances in tissues as they are passed up the food chain.

1489. DDT

Insecticide rarely used because it accumulates in environment.

1490. defoliation

Leaf loss due to natural or artificial agent.

1491. environmental impact statement

A statement of the findings of a detailed study of how an activity might affect the environment.

1492. fungicide

Chemical substance that kills fungi or prevents growth of fungal spores.

1493. green revolution

Using high-yield genetically improved food plants to replace lower-yield indigenous ones.

1494. greenhouse effect

An increase in environmental temperature as carbon dioxide absorbs heat radiated from the earth.

1495. hazardous waste

Environmental pollutants that endanger living things.

1496. herbicide

Chemical substance used to kill plants.

1497. leach

Remove substances by percolation.

1498. nematocide

Chemical substance used to kill nematodes.

1499. nuclear waste

Unused radioactive material from nuclear power plants and other processes involving radiation.

1500. organophosphate

Organic chemical with phosphates attached used in insecticides.

1501. ozone

Molecule consisting of three atoms of oxygen.

1502. ozone shield

A layer of ozone in the upper atmosphere that protects the earth from damaging ultraviolet radiation.

1503. parathion

Organophosphate insecticide.

1504. pesticide

Chemical substance used to kill pests, such as insects.

1505. photochemical smog

Interaction of sunlight and chemical compounds in the air.

1506. plane-sawed

Lumber cut tangentially.

1507. pollution

Detrimental alteration of the environment, usually by human activities.

1508. primary treatment

The first treatment given to sewage.

1509. salinization

Deposition of salt.

1510. secondary treatment

Bacterial decomposition in a sewage treatment plant.

1511. slash-and-burn agriculture

The practice of cutting and burning trees to prepare land for agriculture.

1512. standing crop

Biomass; available food in a habitat.

1513. strip-cropping

The planting of alternating strips of different crops in which one crop protects the other.

1514. sudd

Vegetation broken from plants and floating in a tropical river.

1515. swidden

Temporary agricultural plot made by cutting trees and burning vegetative cover.

1516. temperature inversion

An event that occurs when air near the ground is cooler than air above it.

1517. tertiary treatment

A process that removes toxic substances and excess minerals from sewage effluent.

1518. thermal pollution

Abnormally high temperature produced in an environment.

1519. toxin

Poisonous substance.

Applied Botany

1520. agronomy

Study of agricultural crop production.

1521. allergenic

Able to cause an allergic reaction.

1522. botanical garden

Site of living herbarium where many different plants from many parts of the world grow.

1523. clear-cutting

Felling of all timber in an area.

1524. cleft graft

Graft made by horizontal cut, splitting cut end, and inserting scions so the cambia touch.

1525. cocaine

Narcotic alkaloid from coca leaves.

1526. codeine

Narcotic alkaloid from opium poppy.

1527. compost

Decomposing organic material for use as fertilizer.

1528. contour cultivation

The planting of crops across a slope to slow water runoff.

1529. crop rotation

A system of changing the crop grown in a field to control pests and improve soil fertility.

1530. cull

Separate; removed undesirable organisms or things.

1531. cultigen

Cultivated plant not known to be indigenous to any region.

1532. cultivar

Plant variety under cultivation.

1533. digitalis

Glycoside from leaves of foxglove plant used as heart stimulant in humans.

1534. fodder

Feed for lifestock.

1535. forestry

Study of forest trees and their management.

1536. furocoumarin

Photosensitive substance in some plants (wild parsnips) that can cause blisters and skin inflammation.

1537. grafting

A cloning method by which a dormant stem piece is inserted into another stem with a root system.

1538. hallucinogen

Substance that causes hallucinations.

1539. heroin

A narcotic derived from morphine.

1540. horticulture

Study of garden and orchard plants.

1541. hydroponics

Procedure for growing plants in nutrient solutions without soil.

1542. immune reaction

Production of antibodies or other responses that inactivate or destroy foreign substances.

1543. intercropping

Growing of two or more crops simultaneously in same plot, such as interspersing lettuce and radishes.

1544. knot

Structure in wood representing junction of a branch and the main trunk of a tree.

1545. LSD

Lysergic acid diethylamide, a powerful hallucinogen partly derived from ergot fungus.

1546. lysergic acid

Alkaloid from ergot fungus used to make LSD.

1547. mariculture

Growing of human food in the ocean.

1548. mescaline

Hallucinogenic alkaloid from peyote cactus that induces visions in color.

1549. monoculture

The growing of only one species of crop over a large area.

1550. morphine

Alkaloid from opium poppy.

1551. mulch

Protective cover of bark, straw, or other material used around plants.

1552. nematode

Small, nonsegmented worm widely distributed in soils.

1553. pheromone

Sex-attractant hormone from female insects, which can be used to trap males and control breeding.

1554. polder

A region drained of water and surrounded by dikes.

1555. prune

Cut off superfluous parts.

1556. psilocybin

Hallucinogen in mushrooms of genus Psilocybe.

1557. quarter-sawed

Lumber cut radially and passing through the center of a tree.

1558. quinine

Alkaloid from bark of Cinchona tree used to treat malaria.

1559. reserpine

Alkaloid from snakeroot plant that has been used to treat schizophrenia.

1560. ricin

Toxin from seeds of castor bean.

1561. rootstock

Portion of a root suitable for grafting.

1562. selective cutting

Felling of mature and defective trees, leaving room for young healthy trees to grow.

1563. side graft

Graft in which scion is inserted into the side of a stock plant and left until a union is established.

1564. silviculture

Commercial tree farming.

1565. spring wood

Xylem containing large cells that was produced in early spring.

1566. stock

In grafting, the rooted stem that supports the scion.

1567. summer wood

Xylem containing small cells that was produced in summer.

1568. terracing

Farming on banks of soil built across a slope to reduce erosion.

1569. THC

Tetrahydrocannabinol, the component of marijuana that produces intoxication.

1570. turpentine

Resin distillate used to thin paints and varnishes.

Index

Ascomycetes	95	bog	144
ascospore	108	bole	31
ascus	108	bordered pit	64
asexual	87	bordered pitted tracheid	64
assay	82	boreal	144
assimilation	63	botanical garden	153
association	137	botany	1
aster	49	brackish	144
atom	8	bract	72
atomic number	8	branch	64
atomic weight	8	branch root	64
autecology	1	bread mold	109
autodigestion	12	breathing root	64
autoecious	82	bryology	116
autoploidy	41	Bryophyta	95
autosomal	41	bryophyte	116
autosome	41	bud	64
autotroph	72	bud primordium	54
auxin	82	bud scale	64
auxotrophic	41	budding	87
axial	34	buffer	8
axil	63	bulb	121
axile placentation	121	bundle cap	54
axillary bud	63	bundle sheath	54
bacillus	105	buttress root	64
bactericidal agent	105	C-3 pathway	72
bacteriostatic agent	105	C-4 pathway	72
bacterium	105	calcareous	34
bark	54	callus	54
basal body	19	calorie	72
basal placentation	121	calyx	127
base	8	cambium	54
basidiocarp	109	canopy	144
Basidiomycetes	109	capillary action	64
basidiospore	109	capsule	127
basidium	109	carbohydrate	12
benthos	144	carbon cycle	137
berry	126	carnivorous	137
betacyanin	121	carpel	127
biennial	31	carrageenan	12
binary fission	105	carrier molecule	26
binding site	26	carrying capacity	137
binomial system	99	casparian strip	54
biodegradable	149	cast	100
biogenesis	99	catabolism	78
biological control	149	catalyst	8
biological evolution	99	cation	8
biological magnification	149	catkin	127
biological oxygen demand	137	caulid	116
biology	1	cell	1
biomass	137	cell culture	19
biome	144	cell cycle	49
biosphere	1	cell division	49
biotechnology	49	cell membrane	26
bipinnately compound	72	cell plate	83
bisexual	87	cell theory	1
bisporangiate cone	116	cell wall	26
blade	72	cellulose	12
blue-green algae	106	centriole	19

159

cytology	20	dilated ray	55
cytoplasm	20	dimorphic	35
cytoplasmic streaming	20	dioecious	122
cytosis	26	diplobiontic	113
cytoskeleton	20	diplococcus	106
cytosol	20	diploid	87
cytostome	106	directional selection	100
dark reaction	73	disaccharide	13
data	2	disruptive selection	100
day neutral	83	distal	35
DDT	149	distromatic	35
deamination	79	divergence	101
deciduous	35	diversified habitat	145
decomposer	138	diversity	2
decussate	35	division	96
deductive reasoning	2	DNA	13
defoliation	149	DNA hybridization study	42
dehiscence	128	DNA ligase	50
dehiscence zone	128	DNA polymerase	50
dehiscent	128	DNA replication	50
dehydration	9	dominant allele	42
deletion	42	dormancy	79
denaturation	9	double fertilization	128
dendrochronology	2	drupe	128
dendroid	31	ebb tide	134
denitrification	138	ecological equivalent	101
density-dependent factor	138	ecological niche	138
density-independent factor	138	ecology	2
deoxyribonuclease	79	ecosystem	2
deoxyribonucleic acid (DNA)	13	ecotone	145
derivative	55	ecotype	145
dermal tissue	55	ectomycorrhiza	65
dermatophyte	109	edaphic	134
desert	145	egg	87
desertification	145	egg apparatus	128
desiccation	35	ejectosome	114
detached meristem	55	electron	9
determinate	83	electron micrograph	21
determinate umbel	122	electron transport system	79
detritus	138	electrophoresis	21
Deuteromycetes	96	element	9
development	2	elicitor	83
diaspore	83	embryo	84
diatom	113	embryo sac	84
diatomaceous earth	113	embryogeny	84
dichotomous key	100	embryonic axis	84
dichotomous venation	73	embryophyte	84
dicotyledon	122	emergent	145
dictyostele	65	emergent tree	31
differential centrifugation	21	endergonic	9
differential migration	100	endocarp	55
differential permeability	26	endodermis	55
differentiation	83	endogenous	35
diffuse growth	83	endomycorrhiza	65
diffuse root system	65	endophyte	32
diffusion	26	endoplasmic reticulum	21
digitalis	154	endosperm	122
dihybrid	42	endosperm nucleus	122
dikaryotic	109	endospore	106

161

hydrophyte	32	isotonic	27
hydroponics	155	isotope	10
hydrothermal vent	140	karyogamy	110
hyperosmotic	27	karyotype	45
hypertonic	27	kelp	114
hypha	110	kilocalorie	74
hypocotyl	123	kinetic	10
hypodermis	56	kinetochore	22
hypogynous	123	kingdom	3
hypolimnion	146	knee	66
hyposmotic	27	knot	155
hypothallus	110	Krebs cycle	80
hypothesis	3	lacuna	56
hypotonic	27	lagging strand	51
hypsophyll	123	laminate	37
immune reaction	155	lanceolate	37
imperfect flower	129	lateral	37
imperfect stage	110	lateral bud	66
incomplete dominance	44	laterite	135
incomplete flower	123	lauraceous	96
indehiscent fruit	129	leach	150
indeterminate	123	leading strand	51
indeterminate head	129	leaf axil	85
inducer	50	leaf gall	32
inducible enzyme	51	leaf gap	74
inductive reasoning	3	leaf primordium	57
indusium	118	leaf scar	74
inflorescence	123	leaf sheath	74
infrared	135	leaf trace	74
infrutescence	129	leaf vein	74
ingestion	65	leaflet	74
inheritance	44	lecithin	14
innovation	118	legume	97
inoculum	107	lentic	146
inorganic molecule	14	lenticel	66
insectivore	32	leptoma	118
insertion mutation	44	leucosin	14
integument	129	leukoplast	22
intercalary	36	lichen	110
intercalary meristem	56	ligand	28
intercropping	155	light reaction	74
interfascicular	37	lignin	14
internal	37	linkage	45
internal environment	3	linkage group	45
internode	65	lipase	80
interphase	51	lipid	14
interstitial	3	lipid bilayer	28
intertidal region	146	lipoprotein	15
intracellular	22	liter	3
intrinsic	37	littoral zone	146
inversion	44	liverwort	118
ion	10	lobed leaf	74
ionic bond	3	locus	45
irregular flower	129	long-day plant	85
irruptive growth	140	lotic	146
isogamy	89	LSD	155
isolating mechanism	102	lumen	37
isomer	10	lysergic acid	155
isosmotic	27	lysosome	22

166

167

secondary tissue	60	source	81
secondary treatment	151	spatulate	39
secondary wall	60	speciation	104
secondary xylem	60	species	104
secretion	6	specific heat	11
sedge	98	specificity	17
seed	85	sperm	92
seedling	33	spermatogonium	92
segregation	47	spike	132
selection	47	spindle fiber	53
selection pressure	104	spine	77
selective cutting	156	spiral tracheid	60
selectively permeable	29	spirillum	107
self-pollination	125	spongy mesophyll	60
semiconservative replication	53	spongy parenchyma	61
senescence	86	sporangiophore	92
sepal	131	sporangiospore	92
septate	39	sporangium	112
septum	39	spore	92
seral stage	142	spore mother cell	92
sere	142	sporeling	33
serrate	39	sporic meiosis	92
sessile	76	sporocarp	92
seta	119	sporocyte	92
sex chromosome	47	sporogenesis	92
sexual	91	sporophore	92
sheath	29	sporophyll	93
shoot	68	sporophyte	93
shoot system	68	sporopollenin	132
short-day plant	86	sporulation	93
shrub	33	spring wood	157
side graft	157	squamulose	39
sieve area	68	stamen	132
sieve cell	68	stamen tube	132
sieve plate	68	staminate flower	132
sieve tube	68	standing crop	151
sieve tube element	68	starch	17
silica	68	stele	69
siltation	142	steppe	147
silviculture	157	stereoisomer	17
simple fruit	132	steroid	17
simple leaf	76	stigma	132
simple pistil	132	stipe	69
simple tissue	60	stipule	77
sink	81	stock	157
siphonostele	69	stolon	69
slash-and-burn agriculture	151	stoma	77
sleep movement	69	stomatal chamber	77
smooth ER	24	stomium	120
sodium-potassium pump	30	stratification	142
soil structure	135	stratosphere	136
soil texture	136	strip-cropping	151
sol	25	strobilus	93
solstice	136	stroma	25
solute	30	structural gene	47
solution	30	style	132
solvent	30	suberin	30
soredium	91	suberized	30
sorus	120	sublittoral	147

170

www.ingramcontent.com/pod-product-compliance
Lightning Source LLC
Chambersburg PA
CBHW081121170526
45165CB00008B/2515